MOLECULAR, CELLULAR, AND METABOLIC FUNDAMENTALS OF HUMAN AGING

MOLECULAR, CELLULAR, AND METABOLIC FUNDAMENTALS OF HUMAN AGING

EVANDRO FEI FANG

LINDA HILDEGARD BERGERSEN

BRIAN C. GILMOUR

Academic Press is an imprint of Elsevier
125 London Wall, London EC2Y 5AS, United Kingdom
525 B Street, Suite 1650, San Diego, CA 92101, United States
50 Hampshire Street, 5th Floor, Cambridge, MA 02139, United States
The Boulevard, Langford Lane, Kidlington, Oxford OX5 1GB, United Kingdom

Copyright © 2023 Elsevier Inc. All rights reserved.

No part of this publication may be reproduced or transmitted in any form or by any means, electronic or mechanical, including photocopying, recording, or any information storage and retrieval system, without permission in writing from the publisher. Details on how to seek permission, further information about the Publisher's permissions policies and our arrangements with organizations such as the Copyright Clearance Center and the Copyright Licensing Agency, can be found at our website: www.elsevier.com/permissions.

This book and the individual contributions contained in it are protected under copyright by the Publisher (other than as may be noted herein).

Notices

Knowledge and best practice in this field are constantly changing. As new research and experience broaden our understanding, changes in research methods, professional practices, or medical treatment may become necessary.

Practitioners and researchers must always rely on their own experience and knowledge in evaluating and using any information, methods, compounds, or experiments described herein. In using such information or methods they should be mindful of their own safety and the safety of others, including parties for whom they have a professional responsibility.

To the fullest extent of the law, neither the Publisher nor the authors, contributors, or editors, assume any liability for any injury and/or damage to persons or property as a matter of products liability, negligence or otherwise, or from any use or operation of any methods, products, instructions, or ideas contained in the material herein.

ISBN: 978-0-323-91617-2

For information on all Academic Press publications visit our website at https://www.elsevier.com/books-and-journals

Publisher: Andre G. Wolff
Acquisitions Editor: Michelle W. Fisher
Editorial Project Manager: Kristi Anderson
Production Project Manager: Punithavathy Govindaradjane
Cover Designer: Matthew Limbert

Typeset by TNQ Technologies

Contents

List of contributors	*ix*
Foreword	*xi*
Preface	*xiii*

1. The hallmarks of aging: decoding aging's mystery — 1

Brian C. Gilmour, Linda Hildegard Bergersen and Evandro Fei Fang

1. Aging	1
2. Aging research	2
3. Life span and health span	3
4. The hallmarks of aging	4
References	5

SECTION I The genome

2. Epigenetic aging and its reversal — 9

Cristina de la Parte and Diana Guallar

1. Introduction to epigenetics	9
2. Changes in chromatin structure during aging	14
3. Changes in histones during aging	15
4. Changes in histone posttranslational modifications	19
5. Deoxyribonucleic acid methylation in aging	23
6. Noncoding ribonucleic acids in aging	26
7. Reversal of epigenetic aging	27
8. Conclusions and perspectives	27
References	30

SECTION II Metabolism, homeostasis, and communication

3. Nutrient sensing and aging — 41

Lili Yang

1. AMPK signaling pathway	42
2. mTOR pathway	43

v

vi Contents

3. SIRT pathway	44
4. IGF-1 pathway	45
5. Calorie restriction and aging	46
6. Conclusion	48
References	48

4. Dysregulated proteostasis: mechanisms and links to aging 55

Yasmeen Al-Mufti, Stephen Cranwell and Rahul S. Samant

1. Introduction	55
2. Functional modules of the proteostasis network	56
3. Proteostasis network dysregulation in aging	72
4. Interplay with other aging hallmarks	83
5. Targeting proteostasis for healthy aging	85
6. Concluding remarks	87
Acknowledgments	88
References	88

SECTION III   Autophagy and bioenergetics

5. Autophagy and bioenergetics in aging 107

Jianying Zhang, He-Ling Wang and Evandro Fei Fang

1. Compromised autophagy and mitophagy in aging and disease	107
2. Mitochondrial dysfunction	119
3. Mechanisms of mitochondrial dysfunction	122
4. Dysfunctional mitochondria contribute to energy shortages and aging cells	132
5. Conclusions and future perspectives	136
Acknowledgments	137
References	137

SECTION IV   Senescence

6. Senescence in aging 149

Sofie Lautrup, Alexander Anisimov, Maria Jose Lagartos-Donate and
Evandro Fei Fang

1. What is senescence?	149
2. Current concepts of senescence in aging	159
3. Factors driving and/or triggering senescence in aging	165
4. Stem cell exhaustion in aging	173

Contents **vii**

   5. To develop senotherapies (senolytics and senomorphics) against
age-related diseases and beyond   176

   6. Methods for analyzing and quantifying senescence in vitro and in vivo   177

   7. Concluding remarks   181

   Acknowledgments   181

   References   182

SECTION V   Applications

7. Aging and the immune system   199

Wenliang Pan

   1. Overview of immunosenescence   199

   2. Aging-related lymphoid organs and immunity   199

   3. Interplay between immunosenescence and aging process   215

   4. Conclusions   221

   References   222

8. Canonical and novel strategies to delay or reverse aging   225

Brian C. Gilmour, Linda Hildegard Bergersen and Evandro Fei Fang

   1. Approaching aging as a treatable disease   225

   2. Nonpharmaceutical interventions against aging   226

   3. Pharmaceutical treatments against aging   228

   4. Novel approaches to slow or reverse aging   231

   5. Outstanding questions and future perspectives   234

   Acknowledgments   234

   References   235

Index   *241*

List of contributors

Yasmeen Al-Mufti
Signalling Programme, The Babraham Institute, Cambridge, United Kingdom

Alexander Anisimov
Department of Clinical Molecular Biology, University of Oslo and Akershus University Hospital, Lørenskog, Norway

Linda Hildegard Bergersen
The Norwegian Centre on Healthy Ageing (NO-Age) Network, Oslo, Norway; The Brain and Muscle Energy Group, Electron Microscopy Laboratory, Department of Oral Biology, University of Oslo, Oslo, Norway; Synaptic Neurochemistry and Amino Acid Transporters Labs, Division of Anatomy, Department of Molecular Medicine, Institute of Basic Medical Sciences and Healthy Brain Ageing Centre, University of Oslo, Oslo, Norway; Center for Healthy Ageing, Department of Neuroscience and Pharmacology, Faculty of Health Sciences, University of Copenhagen, Copenhagen, Denmark

Stephen Cranwell
Signalling Programme, The Babraham Institute, Cambridge, United Kingdom

Evandro Fei Fang
Department of Clinical Molecular Biology, University of Oslo and Akershus University Hospital, Lørenskog, Norway; The Norwegian Centre on Healthy Ageing (NO-Age) Network, Oslo, Norway

Brian C. Gilmour
The Norwegian Centre on Healthy Ageing (NO-Age) Network, Oslo, Norway; Department of Molecular Medicine, Institute of Basic Medical Sciences, University of Oslo, Oslo, Norway

Diana Guallar
Universidade de Santiago de Compostela, Center for Research in Molecular Medicine and Chronic Diseases, Santiago de Compostela, Spain

Maria Jose Lagartos-Donate
Department of Clinical Molecular Biology, University of Oslo and Akershus University Hospital, Lørenskog, Norway

Sofie Lautrup
Department of Clinical Molecular Biology, University of Oslo and Akershus University Hospital, Lørenskog, Norway

Wenliang Pan
Division of Rheumatology and Clinical Immunology, Department of Medicine, Beth Israel Deaconess Medical Center, Harvard Medical School, Boston, MA, United States

Cristina de la Parte
Universidade de Santiago de Compostela, Center for Research in Molecular Medicine and Chronic Diseases, Santiago de Compostela, Spain

Rahul S. Samant
Signalling Programme, The Babraham Institute, Cambridge, United Kingdom

He-Ling Wang
Department of Clinical Molecular Biology, University of Oslo and Akershus University Hospital, Lørenskog, Norway

Lili Yang
Department of Nutrition, School of Public Health, Sun Yat-sen University, Guangzhou, China

Jianying Zhang
Department of Clinical Molecular Biology, University of Oslo and Akershus University Hospital, Lørenskog, Norway; Xiangya School of Stomatology, Central South University, Changsha, China

Foreword

In recent years, we have seen tremendous developments in our understanding of the processes of aging and how they may be modified. Thus, it is timely for Dr. Bergersen and Dr. Fang and Mr. Gilmour to have compiled a comprehensive overview of the biology and underlying processes of aging.

In this book, these authors help us to understand the underlying biological processes of aging in a clear and readily understandable manner, as well as some approaches to our lifestyle to achieve healthier aging. The population is rapidly aging, and with this comes many age-associated diseases, which are a huge burden for individuals and society. Even small changes in lifestyle can increase the number of healthy years and delay the consequences of aging. For example, the authors discuss how better sleep can help us attain that goal and how simple modifications in lifestyle and the intake of some natural substances may lead us to achieve more healthy aging.

As the authors point out, much insight into how these lifestyle changes may enable us to achieve more healthy aging has been revealed in recent years, and it has become evident that we can make modifications to our lifestyle that are likely to have a great impact on our future years, even if we begin these changes at an advanced age. It is never too late to commence such changes; even minor ones can have a significant influence. Thus, this is a well-timed volume written by engaged authors, and will enlighten us going forward.

Vilhelm A. Bohr

National Institute on Aging, National Institutes of Health, United States and University of Copenhagen, Denmark

Preface

Sometimes you meet a person who has a lot of energy and enthusiasm, whose ideas cannot be overlooked. Meeting Evandro Fei Fang was this kind of event.

Both of us were interested in everything that had to do with aging and how to prevent it from becoming a bad experience. We already had a lot of science about aging in our suitcase, but how best to communicate it all? What about young people enrolled in different universities worldwide who are interested in aging research? We soon asked the question: what about an antiaging book for nonscientific people, especially student colleagues who are interested in aging? Yes, why not! A wonderful idea. We began on a book intended for the population at large, regardless of scientific literacy. Evandro Fei Fang and I started to design the outline and write the chapters in early 2019. Our dedication and passion did not fade but grew stronger during the long COVID-19 pandemic, as we noted how much the disease disproportionately affected the elderly population. In summer 2020, a young PhD student, Brian C. Gilmour, suggested we pivot our efforts to a scientific textbook devoted to the fundamentals of aging. Thus, this textbook was born, and now we are pleased to have brought this book to publication. We hope this book will inspire antiaging research to equip us all better, especially elderly people, so that in the future, we may respond better to similar outbreaks. Finally, we hope that our book will open the eyes of the younger generation and educate them well, increasing their curiosity and filling them with the same enthusiasm that has driven both of us throughout our work in the field.

In the journey of preparing *Molecular, Cellular, and Metabolic Fundamentals of Human Aging*, we were pleased to have the continued help of Brian C. Gilmour, originally from Canada. Brian has been a news editor of the Norwegian Centre on Healthy Ageing Network (NO-Age; www.noage100.com), cofounded by Evandro Fei Fang, Hilde Nilsen, Jon Storm-Mathisen, and myself. In this book, Brian C. Gilmour contributed to inviting scientists from different aging-related fields to serve as authors for specific chapters. Under our guidance, he also wrote a few chapters and provided proofreading and language editing.

This book covers the molecular mechanisms of several major hallmarks of aging and connects these hallmarks to the physiology of aging and age-predisposed disease. In addition, it details the latest progress on genetic and pharmaceutical approaches that may delay aging.

We thank Prof. Vilhelm A. Bohr, a world-leading scholar on aging and our old friend, in supporting this project. Without Kristi L. Anderson, senior editorial project manager at Elsevier, we could not have had this book published on time.

Linda Hildegard Bergersen
July 9, 2022, Oslo

CHAPTER 1

The hallmarks of aging: decoding aging's mystery

Brian C. Gilmour[1,2], Linda Hildegard Bergersen[1,3,4,5] and Evandro Fei Fang[1,6]

[1]The Norwegian Centre on Healthy Ageing (NO-Age) Network, Oslo, Norway; [2]Department of Molecular Medicine, Institute of Basic Medical Sciences, University of Oslo, Oslo, Norway; [3]The Brain and Muscle Energy Group, Electron Microscopy Laboratory, Department of Oral Biology, University of Oslo, Oslo, Norway; [4]Synaptic Neurochemistry and Amino Acid Transporters Labs, Division of Anatomy, Department of Molecular Medicine, Institute of Basic Medical Sciences and Healthy Brain Ageing Centre, University of Oslo, Oslo, Norway; [5]Center for Healthy Ageing, Department of Neuroscience and Pharmacology, Faculty of Health Sciences, University of Copenhagen, Copenhagen, Denmark; [6]Department of Clinical Molecular Biology, University of Oslo and Akershus University Hospital, Lørenskog, Norway

Consume my heart away; sick with desire

And fastened to a dying animal

It knows not what it is; and gather me

Into the artifice of eternity.

*—**Sailing to Byzantium, W.B. Yeats***

1. Aging

Aside from love, death is one thing for which the entirety of human history has produced no progress, a problem for which "no explanation, no solution, has yet been discovered; [and thus] it will always be impossible to locate a common rule, resting on consensus," to quote Rilke.[1] Whereas any study of death remains largely closed off to science and its methods, great bounds in progress have been made in the condition that inevitably leads to it (i.e., the study of aging). Of course, aging as a concept has long been recognized, producing as it does a ubiquitous and characteristic phenotype: gray hair, wrinkled skin, reduced mobility, loss of hearing, changes to the spinal structure, and an overall decrease in health, among others.

But whereas many aspects of aging have long been understood, this comprehension has progressed little beyond a recognition of the surface conditions, with little gain in knowledge occurring until the advent of modern science, which has begun to put together the minutiae and mechanisms that drive this phenomenon.

Molecular, Cellular, and Metabolic Fundamentals of Human Aging
ISBN 978-0-323-91617-2
https://doi.org/10.1016/B978-0-323-91617-2.00007-9

© 2023 Elsevier Inc.
All rights reserved.

It has long been the prevailing opinion that aging is inevitable and unavoidable, that there can be no cure or treatment for aging because it is a natural phenomenon, a result of our bodies running their course. However, a growing body of research[2-6] suggests that several routes can be taken to bolster our bodies as we age, and thus delay, or, at the very least minimize, the effects of aging with which we are all familiar. These different routes vary in degree and intensity from simple changes in diet and an increase in exercise, to providing more enriched social environments for the elderly, to more complex approaches, such as more recent and expanding trends in diet, exercise regimens, pharmaceutical intervention, and dietary supplementation.[2,3,7,8] Although a cure for death is still unimaginable, we can now easily envision a world in which enhanced age is less of a debilitating factor, where we can enjoy good health and a good physical condition until much later in life. Much remains to be done and understood, but the foundations for further research have already been laid.

2. Aging research

Whereas for much of the developed world the modern era has separated us greatly from diseases that have been with humanity since time immemorial, the COVID-19 pandemic has served as a reminder that we cannot be complacent about health, especially in the elderly, who were at particular risk throughout the pandemic.[9,10]

Exploration into health and aging, and how to improve the two, however, began long before the outbreak of the pandemic. A great deal can be learned about the current difficulties of any field by looking at its history, its *origin*. Science at the turn of the 20th century was greatly augmented by a sudden understanding of pathogens: *Mycobacterium tuberculosis*, *Vibrio cholerae*, *Streptococcus pneumoniae*, *Corynebacterium diphtheriae*, and *Salmonella typhi*, among many others. These bacteria were named from the diseases they had long caused: tuberculosis, cholera, pneumonia, diphtheria, and typhoid fever. Identification of these pathogens as the mediators of their respective diseases enabled the redirection of efforts of control against them, a feat that had been impossible beforehand. It has given us a priceless gift in the form of antibiotics, which today still form a crucial part of the medical apparatus.

Likewise, the ability of our bodies to fight and often completely beat these pathogens laid the grounds that would lead to the development of theories about the immune system through further identification of other pathogen classes (i.e., viruses, bacteria, and parasites, along with other phenomena such as cancers and immune cells) and the beginnings of immunology as a field of study. Before this, centuries of advances in anatomic understanding had already begun piecing together the functions and co-reliance of the body's many organs.

Unfortunately, in the study of aging the edges have been less clear: no one organ of aging, no single pathogen, and no one dysregulated signaling pathway is responsible. This lack of a single culprit has hampered the study of aging. Thus, the study of aging has a great deal in common with the study of cancer, which likewise lacks a predictable initiation that holds true for all cases, even if underlying patterns occur. In fact, identifying these underlying patterns and the genes that most often have key roles and lead to the failure to eliminate tumors that allow the disease to progress have served to strengthen the field.

As with the study of aging, cancer biology as a line of study did not evolve overnight. Rather, it was based on a foundation that brought together the various fields that have a role in cancer or are affected by it: immunology, genetics, metabolomics, pharmacy, cell biology, anatomy, and so on.

The field of aging science has reached a similar point, in which the various preexisting branches of science that touched on aging have united to form a new and distinct field. This textbook aims to summarize current knowledge regarding several of the main fundamentals of the aging process, as exists in the literature. To accomplish this, we build on excellent research done by many different research groups, institutions, and researchers involved in the field, and compile and consolidate their combined research into a single textbook.

3. Life span and health span

This past century has seen a rapid increase in human life spans[11,12] owing to many factors, including the availability of antibiotics to treat bacterial infections, improvements in corrective surgeries, improvements in organ transplants and prosthetics, and the rise of various medications that can be used to lower blood pressure, cholesterol, and so forth. The effect of this can easily be seen in the great expansion of the elderly population in many developed countries around the world,[13] but this expansion has proved to be a burden socially and economically, especially on global health care systems. Although many people in developed countries are living longer, they are not necessarily living longer in good health.[14] The benefits of the past century have produced numerous tools to reduce overall human mortality but have made little progress in reducing the effects of age on the body, its organs, and its function. Thus, the field of aging still needs both further knowledge and innovation so that we can ensure that the elderly age healthily.

Therefore, another metric should be considered. In addition to the life span, we must pay attention to individuals' health spans. Rather than just reducing mortality, we should also find ways to reduce or hinder the negative effects of aging or slow the process itself. This may have additional benefits beyond a healthier elderly population. Aging is a major predisposing factor to a host of diseases, from neurodegenerative diseases such as Alzheimer's or Parkinson's to the many varieties of cancer, and even to bacterial and viral infections. Enhanced age is linked to increased prevalence, increased severity, and increased overall mortality in all of these diseases.[15]

A prime example of this is pneumonia, a persistent if relatively benign disease in the young that in the elderly quickly spirals out of control and becomes life-threatening.[16] One could imagine that if the immune system of the elderly could be made more youthful, many elderly deaths from pneumonia could be avoided without the need for additional treatment. This highlights just one of many potential benefits of treating the central concept of aging, as opposed to its various concurrent side effects. In other words, it may be simpler, more effective, and more economical to treat aging as a whole to produce a healthier elderly population than to treat every age-related disease individually.

The past century produced an impressive array of new treatments and therapies for a variety of diseases and syndromes. These have become increasingly sophisticated with time. However, these drugs and treatments work best in isolation (as a sole regimen of treatment). One of the main issues in treating the elderly is the multitude of overlapping conditions they experience.[17] This often means that elderly patients are on a long list of concurrent regimens in which varying side effects and drug interactions are nearly impossible to ascertain, and which are difficult for the patient to follow.[18]

4. The hallmarks of aging

A foundation does not rise out of nowhere. This textbook builds on prior work undertaken to unite the various fields of aging research into an intelligible whole. Although imperfect,[19] the 2013 *Cell* article "The Hallmarks of Aging," by López-Otín et al.,[6] has been a great inspiration for organizing this textbook's contents (if not the larger aging research field).

It's continuing importance is reflected in the contents of this textbook, which features several of the hallmarks noted, as well as an additional chapter focusing on compromised autophagy, a more recent hallmark.[2]

Moreover, we have further categorized the hallmarks into broader groups based on shared cellular compartments, overlapping functions or areas, and other such similarities, largely to avoid repeating base material that is common among different hallmarks: genes, proteins, cellular processes, and so on.

In this textbook we have provided both details of current knowledge in many fields related to aging and its hallmarks as well as the fundamentals needed to understand how and why each contributes to the aging process or is affected by it. In this way, we hope to build on the excellent ground laid by the many literature reviews and primary research articles that have already added their piece to the aging puzzle, providing those who may not be familiar or acquainted with aging research easier access to the forefront of the field.

References

1. Rilke RM. *Letters to a Young Poet: San Rafael, Calif.* New World Library; 1992. ©1992; 1992.
2. Aman Y, Schmauck-Medina T, Hansen M, et al. Autophagy in healthy aging and disease. *Nat Aging.* 2021;1:634−650.
3. Cunnane SC, Trushina E, Morland C, et al. Brain energy rescue: an emerging therapeutic concept for neurodegenerative disorders of ageing. *Nat Rev Drug Discov.* 2020;19:609−633.
4. Fang EF, Scheibye-Knudsen M, Chua KF, Mattson MP, Croteau DL, Bohr VA. Nuclear DNA damage signalling to mitochondria in ageing. *Nat Rev Mol Cell Biol.* 2016;17:308−321.
5. López-Otín C, Kroemer G. Hallmarks of health. *Cell.* 2021;184:33−63.
6. López-Otín C, Blasco MA, Partridge L, Serrano M, Kroemer G. The hallmarks of aging. *Cell.* 2013;153:1194−1217.
7. de Cabo R, Mattson MP. Effects of intermittent fasting on health, aging, and disease. *N Engl J Med.* 2019;381:2541−2551.
8. Partridge L, Fuentealba M, Kennedy BK. The quest to slow ageing through drug discovery. *Nat Rev Drug Discov.* 2020;19:513−532.
9. Hu B, Guo H, Zhou P, Shi Z-L. Characteristics of SARS-CoV-2 and COVID-19. *Nat Rev Microbiol.* 2021;19:141−154.
10. Cox LS, Bellantuono I, Lord JM, et al. Tackling immunosenescence to improve COVID-19 outcomes and vaccine response in older adults. *Lancet Health Longev.* 2020;1:e55−e57.
11. Fang EF, Scheibye-Knudsen M, Jahn HJ, et al. A research agenda for aging in China in the 21st century. *Ageing Res Rev.* 2015;24:197−205.
12. Fang EF, Xie C, Schenkel JA, et al. A research agenda for ageing in China in the 21st century (2nd edition): focusing on basic and translational research, long-term care, policy and social networks. *Ageing Res Rev.* 2020;64:101174.

13. Roser M, Ortiz-Ospina E, Ritchie H. *Life Expectancy*. 2019 [online] our world in data.
14. Medina L, Sabo S, Vespa J. *Living Longer: Historical and Projected Life Expectancy in the United States, 1960 to 2060*. MD, USA: US Department of Commerce, US Census Bureau Suitland; 2020.
15. Niccoli T, Partridge L. Ageing as a risk factor for disease. *Curr Biol.* 2012;22:R741–R752.
16. Janssens JP, Krause KH. Pneumonia in the very old. *Lancet Infect Dis.* 2004;4:112–124.
17. Divo MJ, Martinez CH, Mannino DM. Ageing and the epidemiology of multi-morbidity. *Eur Respir J.* 2014;44:1055–1068.
18. Montamat SC, Cusack B. Overcoming problems with polypharmacy and drug misuse in the elderly. *Clin Geriatr Med.* 1992;8:143–158.
19. Gems D, de Magalhães JP. The hoverfly and the wasp: a critique of the hallmarks of aging as a paradigm. *Ageing Res Rev.* 2021;70:101407.

SECTION I

The genome

CHAPTER 2

Epigenetic aging and its reversal

Cristina de la Parte and Diana Guallar

Universidade de Santiago de Compostela, Center for Research in Molecular Medicine and Chronic Diseases, Santiago de Compostela, Spain

1. Introduction to epigenetics

Epigenetics refers to all of the molecular mechanisms involved in gene expression regulation that can be inherited and do not involve changes in the sequence of DNA itself. These regulatory mechanisms include the deposition of histone variants in nucleosomes, posttranslational modifications of histone tails, DNA modifications and several noncoding RNAs (ncRNAs), which together participate in regulating gene expression through the control of chromatin structure and accessibility for transcription factors (TFs) and transcriptional regulators.

1.1 Chromatin structure

Chromatin is the form in which DNA is stored in the nucleus. It consists of DNA wrapped around protein histones forming nucleosomes, which further condense into chromatin fibers, which then condense again to form the chromosome. Packaging of DNA into chromatin is important for cellular physiology because it controls DNA accessibility. As a result, it regulates all genomic processes, including DNA replication, transcription, recombination, and DNA damage repair. Chromatin has two main forms: euchromatin and heterochromatin. Euchromatin refers to the more relaxed and open chromatin, which allows gene expression to occur. Heterochromatin is denser and more compact, making DNA less accessible. These two forms are dynamic and highly influenced by epigenetic modifications, TFs and epigenetic complexes, and nucleosome composition. Heterochromatin can be further subdivided into two fractions: facultative heterochromatin, which is chromatin that can adopt open or closed conformations in different contexts, and constitutive heterochromatin, which is permanently condensed.[1] Heterochromatin's characteristic markers are heterochromatin protein 1 (HP1), reduced levels of histone acetylation, and high levels of the repressive marks dimethylation of lysine 9 in histone 3 (H3K9me2) and trimethylation of lysine 9 in histone 3 (H3K9me3) in facultative or constitutive heterochromatin, respectively.[1,2]

Molecular, Cellular, and Metabolic Fundamentals of Human Aging
ISBN 978-0-323-91617-2
https://doi.org/10.1016/B978-0-323-91617-2.00006-7

© 2023 Elsevier Inc.
All rights reserved.

1.2 Histone variants

Histones can be classified into two groups: canonical and replication-independent histones. The expression of canonical histones is cell cycle—dependent, because they are expressed only during the S phase when DNA replication occurs.[3] Replication-independent histones, also known as histone variants, are expressed throughout the cell cycle and in postmitotic cells. Some are widely expressed, whereas others are tissue specific. Histone variants have key roles in maintaining genome integrity[4] and in cell fate decisions relevant to stem cell regulation,[5–8] somatic cell reprogramming (SCR),[9] epithelial-to-mesenchymal transition,[10] proliferation and senescence,[11–15] and DNA damage responses[16] (reviewed in Ferrand et al.[150]), among others. Histone variants diversify nucleosome structure and function, affect nucleosome stability and chromatin compaction, and have an impact on histone posttranslational modifications and recruitment of chromatin-associated proteins, adding a layer to the epigenetic regulation of the genome.

1.3 Histone posttranslational modifications

Histones can undergo posttranslational modifications that affect nucleosome conformation and the chromatin compaction state, modulating gene expression (reviewed in Tessarz and Kouzarides[151]). Posttranslational modifications are deposited on histone tails and comprise a variety of marks from small chemical groups to large peptides. Histone posttranslational modifications that can be found in chromatin include phosphorylation, methylation, acetylation, ubiquitylation, sumoylation, citrullination, deamination, adenosine diphosphate (ADP) ribosylation, and proline isomerization.[17] The most abundant and studied among the wide variety of known modifications are histone methylations, which can occur at lysine or arginine residues, and histone acetylation at lysines. Histone posttranslational modifications affect histone structure, modulating interactions between histones and/or DNA, or interactions with other chromatin proteins affecting chromatin structure and gene expression. Repressive marks refer to histone modifications that cause transcriptional silencing and/or heterochromatin formation, such as histone 3 lysine 27 trimethylation (H3K27me3), histone 3 lysine 9 monomethylation, dimethylation, and trimethylation (H3K9me1-3), and histone 4 lysine 20 trimethylation (H4K20me3).[18,19] In contrast, activating marks promote active transcription and/or chromatin accessibility, and are exemplified by histone 3 lysine 5 trimethylation (H3K4me3) enriched in active promoters, histone 3 lysine 36 trimethylation (H3K36me3) deposited in the core and at the 3' end of active genes, and active enhancers marks such as histone 3 lysine 27 acetylation (H3K27ac), and histone 4 lysine 16 acetylation marks (H4K16ac).[19-21]

Histone posttranslational modifications can have different functions, depending on the residue that is modified and the modification levels (i.e., mono-, di-, or tri-). This is exemplified by H3K9 methylation marks (i.e., H3K9me1, H3K9me2, and H3K9me3). Whereas H3K9me1 is found at euchromatic sites, H3K9me2 appears to correlate with facultative hetero-chromatin, and H3K9me3 is a marker of constitutive heterochromatin.[22] Moreover, the combination of different modifications on the same chromatin regions generates a code used, for example, in early embryonic development to mark bivalent promoters, by combining H3K4me3 and H3K27me3 activating and repressive marks, respectively, thus ensuring that genes under such regulation are kept silent but poised to be activated upon specific cues, and allowing the activation of transcriptional programs in a coordinated and fast manner.[23,24] The reversible nature of histone tail modifications makes it an attractive druggable target for cell fate and/or behavior modulation.

1.4 Deoxyribonucleic acid modifications

The most abundant and studied modification of DNA is methylation of the fifth carbon of cytosine (also termed 5-methylcytosine [5 mC]), which is mainly present at CpG dinucleotides generally associated with repression of gene expression. DNA methylation is a conserved epigenetic mechanism for transcriptional regulation and genome stability. The deposition of a methyl group in a cytosine is mediated by the DNA methyltransferase (DNMT) protein family.[25,26] DNMT1 is responsible for maintaining methylated sites after cell divisions, and DNMT3a and DNMT3b are de novo methyltransferases. Methylation of DNA is generally linked to transcriptional repression by promoting or blocking the binding of transcriptional repressors or activators, respectively, and recruiting enzymes responsible for the catalysis of repressive histone marks. 5 mC can be subjected to iterative oxidations to 5-hydroxymethylcytosine (5hmC), 5-formylcytosine (5 fC), and 5-carboxylcytosine (5caC) by the ten-eleven translocation (TET) protein family.[27-30] Whereas 5 fC and 5caC are present in the mammalian genome at low levels, 5hmC is abundant during embryonic development, from zygotes to blastocysts,[31,32] and is enriched in the brain and bone marrow.[33,34] 5hmC is more abundant and has been shown to be an intermediary mark of DNA demethylation. It has also gained interest as an epigenetic mark.[30,32,35]

1.5 Noncoding ribonucleic acids

In contrast to the traditional belief that only coding genes are transcribed, we know that the vast majority of the human genome is transcribed, generating an enormous pool of ncRNAs.[36,37] ncRNAs can be classified into two groups based on their size: small ncRNAs, with less than 200 nucleotides, and long ncRNAs, which are longer than 200 nucleotides.[38] Among the small ncRNAs we can find the microRNAs (miRNAs), piwi RNAs, and small interfering RNAs. On the other hand, long noncoding RNAs (lncRNAs) include intronic, intergenic, antisense, and overlapping bidirectional transcripts.[39] lncRNAs have been associated with gene expression regulation through co transcriptional and posttranscriptional mechanisms, through their interaction with DNA, RNA, proteins, or other ncRNAs.[40] lncRNAs can regulate chromatin remodeling or transcription, by recruiting histone- and DNA-modifying enzymes to chromatin.[41,42] This is exemplified by the extensively studied role of the lncRNA Xist in promoting epigenetic silencing of one of the two X chromosomes for dosage compensation in placental mammals.[43,44] On the other hand, small ncRNAs can mediate transgenerational epigenetic inheritance for many traits in *Caenorhabditis elegans*, including response to stress, behavior and life span.[45−47] The improvement in RNA- and chromatin-based technologies has allowed a more comprehensive study of ncRNAs in epigenetic regulation, creating great interest in the epigenetic field (reviewed in Duempelmann et al.[152] and in Statello et al. 2021).[48]

1.6 Models of human aging research

Studies of aging in humans are limited by their genetic and environmental diversity as well as their long life span. Thus, most investigations do not perform longitudinal studies on the same patients, but rather pair samples from individuals of the same age who are otherwise heterogeneous. Alternatively, many studies have used progeroid syndromes (PS), which display accelerated aging, as models for studying human aging. PS are rare genetic disorders classified into those caused by perturbations in the nuclear lamina, such as Hutchinson−Gilford progeria syndrome (HGPS) and those arising from mutations that impair DNA replication and repair, such as Werner syndrome (WS).[49,50] HGPS is a rare genetic disorder caused by mutations in the lamin A gene,[51,52] giving rise to a truncated form of the protein called progerin, whose accumulation alters nuclear lamina conformation and affects nuclear architecture, gene expression, genomic instability, and telomere shortening, among others (reviewed by Gonzalo et al., 2017). On the other hand, WS is a recessive genetic syndrome caused by mutations in the WRN gene, which encodes for a RecQ DNA helicase,[53] which leads to altered DNA replication, repair, and stability and a propensity to develop malignancies (reviewed in Oshima et al. [153]). Both HGPS and WS patients display clinical features of aging at an early age, but the diseases are limited because they are pathological and do not fully share underlying mechanisms with normal physiologic aging. Thus, conclusions drawn from their analysis should be examined carefully. On the other hand, human aging is also being modeled in vitro through the ex vivo culture of cells isolated from individuals from different ages, mostly tissue progenitor or stem cells or fibroblasts. Although it is controversial whether and for how long isolated cells retain molecular features of aging,[54] modeling aging in a dish is helpful to identify the cell-intrinsic factors involved in aging. Nevertheless, behavior or aspects related to the immune response and interactions between organs with the passage of time cannot be modeled using this system. Another widely used model of aging is the study of senescent cells in vitro. Senescence is a terminal and stable state of cell growth arrest that occurs in response to different stresses such as DNA damage, telomere dysfunction, and tissue repair, among others.[55] The accumulation of senescent cells promotes organismal dysfunction and aging (reviewed in Di Micco et al.[154]). Senescence has thus been included as one of the nine hallmarks of aging described by Lopez-Otín et al.[56] However, the study of senescence as a model of aging may be not relevant for studying aging in vivo, unless the tissue of interest accumulates senescent cells with age. Finally, many investigations into aging in humans have been based on results previously obtained in other animal models, from yeast to nonhuman primates, finding striking conservations in several key pathways involved in aging, including nutrient sensing and protein homeostasis, involved in regulating longevity (reviewed in Brunet[155]). Together, the results obtained from different animal or in vitro models have helped to push forward the field of aging, in particular the study of epigenetic changes associated with age.

Age-associated changes in histone and DNA modifications, chromatin states, nucleosome positioning, and ncRNAs are thought to be important regulators of the aging process and thus are regarded as aging hallmarks.[56] In this chapter, we will discuss epigenetic changes occurring during human aging and how they correlate with altered gene expression that can exacerbate the aging process. Given the reversibility of epigenetic modifications, deciphering epigenetic changes that appear with the passage of time and the epigenetic mechanisms that underlie the aging process in the human

2. Changes in chromatin structure during aging

2.1 Age-associated loss of heterochromatin

Both physiologic and premature aging are characterized by nuclear architecture defects, accumulation of DNA damage foci, and loss of heterochromatin.[57-59] Heterochromatin-specific marker H3K9me3 is significantly reduced in cells derived from older donors, along with HP1γ and lamina-associated protein 2α.[58,59] Besides this loss of heterochromatin with age, nuclei from older donors show morphologic, structural, and functional abnormalities.[58,59] Furthermore, aged cells show higher levels of DNA damage, determined by the presence of foci containing the phosphorylated form of the histone H2AX (γ-H2AX), and increased levels of mitochondrial reactive oxygen species.[59] DNA damage seems to contribute to the loss of heterochromatin,[60] and loss of constitutive heterochromatin has been observed in both physiologic aging[58,59] and HGPS[57] and WS[61] premature aging. PS have been critical in demonstrating how the alteration of heterochromatin structure and loss of heterochromatin marks could be associated with accelerated aging. On the other hand, senescent cells feature the appearance of domains with increased heterochromatin, referred to as senescence-associated heterochromatin foci (SAHFs) marked by HP1 and H3K9me3 together with accumulation of the H2A variant macroH2A, which will be discussed later in this chapter.[62,63] More than 30% of chromatin in senescent cells undergoes remodeling, with changes in both activating (i.e., H3K4me3) and repressive (i.e., H3K27me3) histone marks that affect chromatin organization, and finally gene expression.[64,65] Remodeling of repressive marks in senescent cells has been associated with three-dimensional genome reconfiguration, aberrant expression of repetitive elements, and gene expression leakage.[66] Future studies using whole-genome chromosome conformation capture techniques will shed light on the dynamics of chromatin during physiologic aging as well as their impact on transcriptional regulation.

2.2 Changes in chromatin accessibility and nucleosome remodeling

The nucleosome composition and localization in the genome have a great impact on DNA processes, including the regulation of gene expression.[67] Chromatin accessibility, analyzed by Assay for Transposase-Accessible

Chromatin sequencing (ATAC-seq), is also altered during aging. Some regions lose openness and others are more open and accessible with age,[68,69] in line with the loss of heterochromatin discussed earlier. In peripheral blood mononuclear cells, an epigenomic signature of aging has been identified.[69] This signature includes chromatin closing at promoters and enhancers related to the immune functions of the cells, and a more subject-specific chromatin opening at some repressed sites.[69] Ucar et al. found a correlation with changes in chromatin accessibility and in expression levels. The genes at the opening sites were more transcribed, whereas those sites with a denser state showed a decline in expression. In another study, ATAC-seq of CD8+ T cells revealed a loss of openness at promoters and gains at enhancers during aging that were linked to decreased levels of the TF NRF1, whose activity is responsible for maintaining chromatin openness in young cells.[68]

Nucleosomes, the building blocks of chromatin, are formed by DNA wrapped around an octamer of core of histones with histone H1, linking adjacent nucleosomes. Whereas interactions among histones in the octamer are tight and stable, its structure is not static; nucleosomes are highly dynamic in both conformation and composition. Nucleosome remodeling refers to the modification of nucleosome position on DNA or the substitution of canonical histones by histone variants (reviewed in more detail in the next section). It is performed by adenosine triphosphate (ATP)-dependent remodelers. Nucleosome remodeling proteins collaborate with histone modifiers and chaperones for histone editing or removal, to modify the distribution of nucleosomes or alter their composition.[70,71] Nucleosome remodeling and deacetylase (NURD) complex components RBBP4, RBBP7, and HDAC1 were shown to be downregulated in both physiologic and progeroid models (HGPS) of aging.[72] In particular, the authors showed that the downregulation of these NURD components in both systems reduced pericentromeric heterochromatin foci and increased DNA damage.[72] On the other hand, nucleosome remodeling complex SWI/SNF was also shown to promote longevity in worms.[73] More studies will be needed to uncover nucleosome remodeling mechanisms implicated in human aging and their role in regulating the aging process.

3. Changes in histones during aging

Alteration of histone protein levels, both canonical and variants, have been found to be altered with the passage of time. This suggests a potential link between the deregulation of these proteins and the aging process.

3.1 Changes in histone expression levels

Whereas reduced nucleosome occupancy and histone biosynthesis have been reported in other species (from yeast and flies to mice) with aging,[74–76] in humans a downregulation of canonical histones in in vitro cultures of human fibroblasts was observed, but no change in histone levels in cells directly analyzed from patients.[77,78] In a replicative aging model consisting of human fibroblasts maintained in culture until late passages but without becoming senescent, lower expression levels of histones H3 and H4 were reported, compared with early-passage ones.[77] This reduced H3 and H4 biosynthesis caused chromatin changes and redistribution of histone posttranslational modifications. In line with this observation, a reduction in the abundance of canonical histone levels was observed in senescent cells.[79] In contrast, Cheung and collaborators directly used primary immune cells obtained from young (aged <25 years) and older (aged >65 years) individuals. The researchers did not observe a significant change in the expression of canonical histones H3 and H4, with the exception of the central memory CD8[+] T cells, which showed a marked decrease in H3 and H4 histone levels.[78] These data suggest that aging-associated histone loss may occur in a cell type—specific manner or that in vitro maintenance of cells can induce a nonphysiologic reduction in canonical histone levels. An analysis similar to the one performed by Cheung et al., using freshly obtained fibroblasts and other cellular types of interest from old donors could address the physiologic relevance of the findings obtained in vitro by O'Sullivan and Ivanov.

3.2 Changes in histone variants

Histone variants perform specialized functions in the genome, such as chromosome segregation, DNA repair, and regulation of transcription initiation through different structural properties compared with canonical ones.[80] Here, we compile several studies that have shown differences in histone variants (mostly with H2A and H3) in the context of aging.

3.2.1 Histone 2A variants

H2A histone variants H2AX, H2AJ, and mH2A have been linked to senescence and aging (Table 2.1). H2AX variant has an additional C-terminal tail of 13 amino acids longer than canonical H2A, which contains a motif that can be phosphorylated in response to DNA damage.[81] γ-H2AX accumulates in foci and initiates DNA damage signaling, promoting the recruitment of DNA repair complexes to chromatin.[82]

Table 2.1 Histone variants and their changes in human aging.

Histone	Name of variant	Change with age	References
Histone 2A	H2AX	Phosphorylated H2AX is increased in senescence and in cells from aged individuals compared with young ones.	(Scaffidi and Misteli[57,58]; Miller et al.[59])
	H2AJ	Accumulates in senescence and aging	(Contrepois et al.[85])
	MacroH2A	Enriched in heterochromatin in senescent cells	(Douet et al.[90]; Zhang et al.[61])
Histone 3	H3.3	Increased from birth to age 10 years; maintained at high levels until death	(Maze et al.[91])
	CENP-A	Decreased with age in islet pancreatic cells	(Lee et al.[95])

Although one study reported reduced H2AX histone variant levels with extensive passaging of human lung embryonic fibroblasts in vitro,[83] through histone fraction resolution in gels, most later studies have shown an increase in γ-H2AX in similar in vitro systems. Specifically, in senescent human fibroblasts, persistent DNA damaged foci and increased γ-H2AX levels are observed.[84] Moreover, γ-H2AX foci are increased in primary fibroblasts from donors of older age or in HGPS patients, compared with their respective controls.[57–59]

H2AJ, another H2A histone variant, differs from canonical H2A in a single amino acid (one valine is replaced by alanine). It has also been shown to accumulate in vivo and in vitro with aging.[85] Whereas H2AJ is present at low levels in proliferating cells, it accumulates in senescent fibroblasts, increasing in levels up to 10-fold, where its deposition is associated with the expression of proinflammatory genes.[85] Furthermore, H2AJ reveals age-dependent accumulation in human skin cells,[85,86] which supports the in vivo relevance of H2AJ accumulation.

MacroH2A (mH2A) histone is characterized by the presence of a 30-kDa nonhistone domain (macrodomain) at its C-terminus.[87] mH2A is implicated in transcriptional repression through chromatin condensation,[88,89] inhibition of transcriptional initiation, and histone acetylation, which prevent chromatin relaxation (Doyen et al. 2006). In vitro, mH2A is enriched in heterochromatin[90] and at SAHF[62] of senescent cells, and is increased in human fibroblasts with replicative senescence (Kreiling et al.

2011). In vivo, it is accumulated with age in human and mouse livers (Borghesan et al. 2016; Kreiling, 2011) as well as in several other tissues of mouse and baboon origin (Kreiling et al. 2011). This points to macroH2A accumulation as a potential aging biomarker.

3.2.2 Histone 3 variants

The abundance of the histone H3 variant H3.3 has been found to be increased in senescent human fibroblasts[14] as well as during physiologic aging in mice.[91] In senescent human fibroblasts, H3.3 colocalizes with the transcription activating histone modification H4K16ac at promoters.[15] Upon knockdown of the chaperone involved in H3.3 deposition (i.e., HIRA),[15] its decrease in chromatin caused a reduction in H416ac levels. These results suggest a role for H3.3 in maintaining the H4K16ac mark at active promoters and regulating gene expression in senescence. Furthermore, H3.3 has been linked to E2F-mediated transcriptional regulation of senescence, specifically through collaborative repression of cell-cycle coding genes.[14] The N-terminal cleaved version of H3.3, H3.3cs1, causes a loss of transcription activating mark H3K4me3 at E2F target genes, promoting their silencing and cellular senescence (Duarte et al.[14]). This histone variant also increases with population doublings (in vitro aging) in human lung embryonic fibroblasts.[83] These results suggest that H3.3 deposition alters histone posttranslational modifications of landscape, gene expression, and chromatin organization.[92] In human brains postmortem, H3.3 protein levels have been observed to increase gradually from birth to age 10 years. Afterward, H3.3 levels are maintained stable, representing 93% of the total H3 in individuals from age 14 to 72 years.[91] Postmitotic tissues such as the brain have a low turnover of histones, favoring accumulation of histone variants, which could contribute to changes in histone modifications, promoting the chromatin functional decline associated with aging. Because H3.3 is associated more with active epigenetic marks compared with H3 canonical histone (reviewed in Szenker et al.[92]), the increase in H3.3 variant with age could contribute to the decrease in heterochromatin observed with aging.

On the other hand, centromeric protein A (CENP-A) is another histone variant of H3 involved in centromere formation, and thus chromosome segregation during cell division.[93] Reduction in CENP-A has been shown to induce cellular senescence to prevent centromere decondensation and chromosomal loss.[94] CENP-A protein levels are decreased with age in islet pancreatic cells, causing defects in cell proliferation capacity.[95]

Furthermore, in T-cell subpopulations with low CENP-A levels, CENP-A was resynthesized, increasing its levels upon T-cell activation, and cells recovered their proliferative capacity. This suggests a role for CENP-A in regulating the capacity for proliferation that is lost with age.[96]

Table 2.1 lists histone variants and their changes in human aging.

Although an accumulating amount of evidence shows the relevant role of histone variants in human aging, most studies in this respect were performed using animal models such as mice, rats, and baboons. The development of single-cell chromatin analysis techniques will undoubtedly be important for addressing more comprehensively how histone variants and modifications are altered in human aging samples and whether and how this correlates with transcriptional changes.

4. Changes in histone posttranslational modifications

Aging is associated with an increase in activating marks and a decrease in repressive marks in histones, which together contribute to the altered transcriptional profiles associated with the passage of time (reviewed in Sen et al.[156]; Benayoun et al.[157]). Histone modifications influence transcription as well as other genomic processes, such as DNA replication and DNA damage repair[97,98] by regulating chromatin accessibility.

Cheung and collaborators developed an atlas of 40 different chromatin marks in 20 immune human cell types during aging, using samples from both male and female young (aged less than 25 years) and older individuals (greater than age 65 years).[78] Principal component analysis of the data revealed clustering of the epigenetic profiles by age, with no effect of sex. They observed that chromatin modification profiles were more homogeneous in young samples, which suggests that aging causes increased cell-to-cell and interindividual heterogeneity in epigenomic profiles, which could explain similar observations at the transcriptomic level.[99,100] The levels of a wide range of the histone marks analyzed in that study were upregulated in older donors compared with younger ones in most of the cell types analyzed. This suggests that the changes could originate in the hematopoietic precursors or stem cells from which they were derived. A remarkable exception was central memory CD8$^+$ T cells, which had 38 of 40 marks reduced with age. This highlights the importance of generating comprehensive studies able to design tissue-specific antiaging therapies. That study also found increased cell-to-cell variability in histone modifications with age, especially polycomb-repressive complex (PRC)-mediated marks H3K27me3, H3K27me2, and H2AK119Ub. Furthermore, H3K27me3 repressive mark decorated genes were associated with higher transcriptional variability with age, compared with those decorated with the active promoter-enriched H3K4me3 mark. The epigenetic variability observed with aging results in alteration of chromatin states between cells, causing greater transcriptional noise.

In contrast to the epigenomic-wide characterization of chromatin changes in aging, most studies have focused their efforts on understanding changes in specific marks, mainly in histone methylation and acetylation.

4.1 Histone methylation

The methylation of histone tails generates a high diversity of histone modifications with different impacts on the epigenome, depending on the histone and the residue that harbor the methyl group. Histone methylation marks can be classified into activating marks, which promote gene expression by facilitating chromatin openness, and repressive marks that compact chromatin blocking transcription.

In HGPS and WS progeroid models, H3K27me3 levels are decreased,[61,101,102] whereas in senescent human fibroblasts, although there is a local decrease in this mark, a global increase has been documented.[65] On the other hand, the activating mark H3K4me3 has been found to be increased or decreased in a locus-specific manner in senescent human fibroblasts.[65] In contrast, in physiologic aging in immune cells, H3K4me3 levels do not show significant alterations,[78] and they are not changed in WS (Zhang et al.[61]). H3K9me1 is found at higher levels in senescent human fibroblasts, in sharp contrast to heterochromatin marks H3K9me2 and H3K9me3, which are decreased with senescence.[77] Furthermore, H3K9me3 decreases in normal aging, as well as in HGPS and WS (Zhang et al.[61]; Shumaker et al.[101]) and physiologic aging.[58] The decrease in the heterochromatin histone mark H3K9me3 is correlated with a global loss of heterochromatin that occurs in physiologic aging and in PS,[103,104] discussed later in this chapter. Finally, H4K20me2, a mark that promotes genomic stability and is implicated in DNA repair mechanisms, is increased in senescent human fibroblasts, in contrast to the constitutive heterochromatin mark H4K20me3, which is enriched in and is decreased with senescence in human fibroblasts.[77] These observations highlight the connection between changes in histone methylation and the aging process. There is a greater loss of repressive marks rather than a net gain in activating ones, which contributes to the loss of heterochromatin observed with aging.

4.2 Histone acetylation

Acetylation of lysines removes the positive charge of histone tails, reducing interactions between nucleosomes and DNA and allowing chromatin to decondense, making it more accessible for gene activation. Histone acetylation mark H3K56ac is decreased in senescent human fibroblasts,[77] whereas H4K16ac and H3K9ac marks are increased in senescent cells.[15,77] Histone acetylation levels have been linked to life span in many models from yeast to humans.[105] Spermidine is a small molecule that inhibits histone acetyltransferases, causing histone hypoacetylation. With age, spermidine levels decrease, and it has been proven that spermidine supplementation increased life span in all tested models, including human cells.[105]

SIRT1, a member of the sirtuin family of proteins, is a histone deacetylase whose protein levels decrease with the passage of human lung fibroblasts in vitro, owing to posttranscriptional changes.[106] Sasaki and collaborators found that SIRT1 levels in mice are linked to aging, because loss of mitotic activity causes a loss of deacetylase, which correlates with results observed in in vitro mouse embryonic fibroblasts cultures. Histone acetylation and sirtuins have been explored in model organisms (i.e., yeast, flies, mice, and rats), but their role in human aging remains to be elucidated.

Table 2.2 lists all histone posttranslational modifications discussed earlier, as well as their function and changes in human aging.

Table 2.2 Histone posttranslational modifications studied in human aging, their function in chromatin, and change they undergo in human aging.

Histone mark	Function	Change in human aging	References
H3K9me1	Euchromatin in gene bodies, transcriptional repression	Increased in senescent fibroblasts	(O'Sullivan et al. [77])
H3K9me2	Facultative heterochromatin, at gene bodies, transcriptional repression	Decreased in senescent fibroblasts	(O'Sullivan et al. [77])
H3K9me3	Heterochromatin, transcriptional repression	Decreased in senescent fibroblasts and in	(O'Sullivan et al. [77]; Zhang et al.[61]; Shumaker et al.[101];

Continued

Table 2.2 Histone posttranslational modifications studied in human aging, their function in chromatin, and change they undergo in human aging.—cont'd

Histone mark	Function	Change in human aging	References
		aging (progeria and physiologic aging)	Scaffidi and Misteli[58])
H3K27me3	Euchromatin gene bodies transcriptional repression	In senescent fibroblasts undergoes remodeling, in progeroid syndromes decrease and remodeling	(Shah et al.[65]; Zhang et al.[61]; Shumaker et al.[101]; McCord et al.[102])
H4K20me2	Genomic stability, DNA repair	Increased in senescent fibroblasts	(O'Sullivan et al. [77])
H4K20me3	Enriched in heterochromatin	Decreased in senescent fibroblasts, increased in Hutchinson—Gilford progeria syndrome	(O'Sullivan et al. [77]; Shumaker et al.[101])
H3K4me3	Euchromatin, involved in transcriptional activation	Werner syndrome no change; in senescent fibroblasts undergoes remodeling. In physiologic aging no significant alteration.	(Zhang et al.[61]; Shah et al.[65]; Cheung et al.[78])
H3K56Ac	DNA replication and DNA damage response	Decreased in senescent fibroblasts	(O'Sullivan et al. [77])
H4K16Ac	Transcriptional activation and repression	Increased in senescent fibroblasts	(O'Sullivan et al. [77]; Rai et al.[15])
H3K9Ac	DNA damage response and replicative stress	Increased in senescent fibroblasts	(O'Sullivan et al. [77])

Although several studies in model organisms proved that modulation of histone modifications has an impact on life span, either increasing or decreasing it, depending on the mark and the loci (e.g., reduction in H3K4me3 levels increased life span in *C. elegans*[107]), the effect of histone modification on human aging is still unknown. Future studies will be needed to elucidate this promising therapeutical target for human life span or health span expansion.

5. Deoxyribonucleic acid methylation in aging

5.1 Deoxyribonucleic acid methylation changes with age

During the aging process, DNA methylation of the epigenome undergoes remodeling, with global hypomethylation accompanied by local hypermethylation at specific genomic regions (e.g., promoter regions and tumor suppressor genes).[108] In vivo studies of human tissues also found a decrease in methylation in CpG islands located outside promoters but hypermethylation at those located near promoters.[109,110] DNA hypermethylation is associated with transcriptional repression, because the methylated state of cytosines in the genome prevents TFs from binding the DNA.[111,112] Furthermore, methylcytosine-binding proteins are known to recruit histone deacetylases, contributing to transcriptional repression.[111]

Changes in DNA methylation during aging are heterogeneous between tissues and cell types. Human senescent cells show decreased levels of 5 mC compared with control cells.[113] In human blood cells, a 2.4% decrease in 5 mC levels in aged individuals compared with newborns was observed.[114] Changes in the methylome of the cell during aging alter chromatin architecture, derepressing heterochromatin and TF activity, leading to gene expression dysregulation. In human monocytes, both hypomethylated and hypermethylated regions have been described with aging.[115] The hypermethylated regions were mainly found at CpG islands, first exons and inactive chromatin, whereas regions with hypomethylation compared with younger ones were found at CpG island shores and 3′ UTR regions, which were predicted also to be enriched for TF binding sites.[115] The researchers also found that a small fraction of differentially methylated sites with age was associated with differential expression of nearby genes, indicating that age-associated changes in the methylome may affect gene expression in human monocytes. A study comparing DNA methylation profiles of human monocytes from young (aged 24−30 years) and older individuals (aged 57−70 years) revealed a subset of age-associated differentially methylated

24   Molecular, Cellular, and Metabolic Fundamentals of Human Aging

regions and a cell type—specific aging signature in the DNA methylation profiles.[116] Furthermore, the sites with increased methylation with age were associated with H3K27me3 repressive marks in CpGs near promoters of genes with low expression, whereas hypomethylated regions were associated with the active enhancer marker H3K4me1 at genes with age-associated increased expression. Future studies are required to address the hierarchy of the distinct epigenetic players (i.e., histone and DNA modifications) during aging, to design efficient prolongevity epigenetic-based strategies.

5.2 Methylation clocks and predictors of health span and lifespan

Taking advantage of these changes in the methylated state of CpGs, which occurs with aging, two pioneer studies developed tools to predict the chronologic age of an individual according to methylation status, based on a small number of CpG sites.[117,118] Bocklandt and collaborators developed a tool to predict age from saliva, with an accuracy of ±5.2 years. Although it was able to predict age from various cell types, the Koch and Wagner clock had a larger error (±11 years). One year later, two other DNA methylation clocks based on a higher number of CpGs were reported to be more accurate.[119,120] Although these later tools possess higher accuracy in predicting chronologic age, with a correlation value higher than 0.9 and median errors between predicted and actual age inferior to 5 years, they present several differences. Horvath's methylation clock uses 353 sites, 193 with methylation positively correlated with age and the other 160 negatively correlated with age. The clock of Hannum et al. is based on 71 CpG sites. Both epigenetic clocks only have six CpGs in common. Another difference between them is that the Horvath clock was developed using data from multiple tissues, which makes it perform well with a variety of samples, whereas the clock of Hannum et al., which was developed using only blood methylome data, needs to be adjusted for use with other sample types. The Horvath clock can monitor methylation modifications that are independent of cell proliferation, and is able to predict the epigenetic age of the brain. Horvath and collaborators did not see the correlation between age-associated changes in methylation and age-associated changes in mRNA levels, suggesting that observed changes in methylation are not directly linked to changes in gene expression (Horvath[119]). However, Hannum et al. detected variations in methylation associated with gene expression changes occurring with age.[120] Despite the high correlation coefficients of these epigenetic clocks, there are significant deviations between predicted and chronologic age. This suggests that methylation profiles may also reflect the biological age of the individuals, which in some cases do not match their chronologic age. DNA methylation clocks can also be used to predict the health of tissues. As an example, in livers from obese patients, an increased epigenetic age compared with the chronologic age of the patient and the epigenetic age of other tissues was observed[121]. A comprehensive review of epigenetic clocks can be found in Simpson et al.[122]

In many cases, chronologic age (years since the individual was born) does not match epigenetic age (based on the methylation clock). Life span is defined as the length of time a person lives, and health span is the period of life spent in good health, free of diseases and disabilities associated with aging. To generate new and powerful epigenetic biomarkers of health span and life span, clinical predictors were incorporated into the epigenetic clocks. One example is the time-to-death predictor developed by Zhang and collaborators,[123] which identified epigenome-wide DNA methylation signatures related to mortality. They reported 58 CpGs, 38 of which are disease-related genes whose methylation is strongly associated with mortality. Another predictor is DNAm PhenoAge, which is able to predict life span by regressing a phenotypic measure of mortality risk on CpGs. It is based on nine clinical blood markers, the chronologic age of the patient, and the methylation state of 513 CpGs.[124] Increased epigenetic age relative to chronologic age is associated with higher levels of proinflammatory cytokines and decreased transcription and DNA damage responses. The most recent life span and health span predictor based on DNA methylation is GrimAge.[125] This predictor estimates the time to death, time to coronary disease, or time to cancer, based on a collection of markers that includes DNA methylation, the plasma levels of seven proteins, the duration that the patient has been a smoker, the patient's chronologic age, and sex. Based on DNA methylation, Lu and collaborators developed a tool able to predict telomere length in leukocytes, called DNAm Telomere Length.[126] This can be used to estimate life span or age-related pathologies (time to coronary disease or congestive heart failure).

The Dunedin Study carried out by Belsky[127] developed a measure of pace of biological aging in individuals based on 18 biomarkers tracking organ-system integrity over 12 years. They developed a DNA methylation clock able to predict the pace of aging, called DunedinPoAm, a powerful tool able to predict the pace of aging of an individual based on a single blood test.

But, when does aging really begin? Remarkably, using specific methylation sites on DNA and seven different reported DNA methylation clocks, a study showed that developing embryos already display markers of epigenetic aging.[128] In particular, Kerepesi and colleagues reported that epigenetic age decreases during the first stages of early embryogenesis, reaching the youngest state, or epigenetic ground zero at embryonic stage E10.5 in mice.[128] From that point onward, the epigenetic age of the embryos increased in both murine and human embryos, suggesting the intriguing possibility that aging starts at early embryonic stages, even before we are born.

6. Noncoding ribonucleic acids in aging

Changes in the expression of nine miRNAs were found to be present in lower levels in older humans compared with young controls.[129] In line with this, the targets of those miRNAs were upregulated with age, supporting the potential role of these RNAs in physiologic aging, although the direct connection between those events still needs experimental validation.

On the other hand, the lncRNA H19 has been correlated with the aging process in mice and with coronary diseases in humans through regulation of the STAT3 pathway.[130] H19 expression is decreased during aging and in cardiovascular disease (atherosclerotic plaques), causing a reduction in the function of endothelial cells. A decrease in H19 also causes an increase in inflammatory activation, resulting in aging-associated chronic inflammation (inflammaging).

Furthermore, a study identified a set of aging-associated lncRNAs that are highly conserved among species, from invertebrates such as worms to higher eukaryotes such as humans.[131] This evolutionary conservation suggests an important regulatory role for these lncRNAs. That study showed significant differences between aging-associated lncRNAs and non—aging associated lncRNAs, because aging-associated lncRNAs contain a higher number of TF binding sites, suggesting a strong function for lncRNAs in regulating TF function. Aging-associated lncRNAs also have a higher number of RNA-binding protein binding sites that regulate RNA maturation, and lower content in transposable elements than non—aging associated lncRNAs. Importantly, aging-associated lncRNAs regulate and are regulated by the nuclear factor-κB signaling pathway, which is known to be important in inflammaging.

ncRNAs are emerging as regulators of many physiologic processes, including aging, and their study will help to decipher molecular mechanisms underlying their contribution to the aging process. Future studies will be needed to determine all changes in ncRNAs that occur with age and how they affect the aging process, uncovering new potential therapeutical targets.

7. Reversal of epigenetic aging

SCR consists of the dedifferentiation of an adult somatic cell into an embryonic-like stem cell. The generation of induced pluripotent stem cells (iPSCs) from somatic fibroblasts was first achieved by overexpression of four transcription factors considered to be master regulators of pluripotency.[132,133] During reprogramming, most aging hallmarks are erased, including epigenetic modification alteration.[134] Epigenetic marks such as DNA methylation, histone modification, and chromatin structure undergo remodeling throughout SCR.[135] The reduced H3K9me3 levels observed in old fibroblasts were restored to younger levels not only in iPSCs after reprogramming but also upon redifferentiation of these pluripotent cells to fibroblasts, demonstrating rejuvenation of the epigenome through reprogramming,[59] Moreover, the introduction of reprogramming factors into senescent human fibroblasts showed increased mobility of the epigenetic factor HP1β[136] which was reset to levels similar to those found in young fibroblasts, further suggesting epigenetic rejuvenation of the cells by SCR. A rising number of studies are reporting the partial reprogramming of human cells as a promising tool for reversing aging features without losing cell identity.[137−140] Epigenetic rejuvenation through partial reprogramming is blocked upon knockdown of TET1 or TET2 DNA demethylases, suggesting the importance of DNA methylation profiles in cellular rejuvenation.[141] In that study, Lu et al. also observed alterations in DNA methylation levels upon tissue damage, which were restored after partial reprogramming in mice. Although reprogramming through overexpression of transcription factors could not be used in clinics as it is, the findings derived from those studies unravel epigenetic pathways and targets whose manipulation could be promising in regenerative medicine. A partial reversal of the epigenetic age (i.e., 1.5 years in epigenetic age) has also been achieved through pharmacologic intervention, through the administration of drugs to enhance thymic function, such as recombinant human growth hormone, dehydroepiandrosterone (DHEA), and metformin.[142]

8. Conclusions and perspectives

As described in this chapter, epigenetic regulation encompasses multiple interconnected mechanisms forming a complex network that regulates nuclear organization and chromatin structure and function. During human aging, accumulation of histone variants, loss and redistribution of nucleosomes, and reduction of heterochromatin result in an altered nuclear architecture. In addition, histone posttranslational modifications and DNA methylation changes observed with age result in a completely different nuclear and genomic state in cells from aged individuals, compared with their younger counterparts. Altogether, these epigenetic modifications that appear in the nucleus can give rise to changes in gene expression, affecting cellular physiology and homeostasis and thus contributing to the molecular mechanisms driving aging (Fig. 2.1).

Epigenetic studies of human aging have historically focused on identifying DNA and histone modifications, but as discussed in this chapter, many more epigenetic mechanisms have important roles in nuclear function and thus could contribute to aging. On the one hand, little is known about the implication of ncRNAs in gene expression regulation and their role in the aging process., RNA has been described as forming a new layer of transcriptional regulation in the cell, effecting change through epigenetic modulation,[143,144] and thus the study of the contribution of RNA to

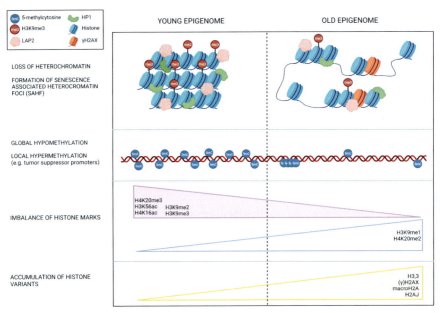

Figure 2.1 Main epigenetic alterations observed in humans during aging. Created with BioRender.com. *5mC*, 5-methylcytosine; *γH2AX*, phosphorylated histone H2AX; *H3K9me1*, histone 3 lysine 9; monomethylation; *H3K9me2*, histone 3 lysine 9 dimethylation; *H3K9me3*, histone 3 lysine 9 trimethylation; *H4K16ac*, histone 4 lysine 16 acetylation; *H4K20me2*, histone 4 lysine 20 dimethylation; *H4K20me3*, histone 4 lysine 20 trimethylation; *HP1*, heterochromatin protein 1; *LAP2*, lamina-associated protein 2; *SAHF*, senescence-associated heterochromatin foci.

epigenetics will be important to a further understanding of the aging process. As well, the study of RNA chemical modifications (i.e., epitranscriptomics) has been noted as a critical determinant for histone modifications in chromatin,[48,145,146] but their contribution to the drifting epigenome configuration observed during aging must still be clarified.

Most studies analyzing methylation of DNA during aging, including to define the methylation clocks, use reference standard bisulfite conversion technologies and sequencing, which cannot distinguish between methylated cytosines and other DNA modifications such as 5hmC. Although less abundant than 5 mC, 5hmC can be readily detectable at considerable levels in several tissues,[33,34] where BS-seq would not reliably discriminate 5 mC from 5hmC-modified cytosines. Using mass spectrometry, a study determined 5hmC levels in human blood cells with age, and reported a 27.5% decrease from birth to geriatric age.[114] This reduction, which occurs progressively with the passage of time, suggests that 5hmC could be used to develop epigenetic clocks based on the presence of this mark in the epigenome. Future studies specifically discriminating 5 mC from 5hmC using chemical conversion (i.e., oxidative bisulfite sequencing (oxBS-seq) in combination with bisulfite sequencing (BS-seq)) or antibody-based methods (i.e., 5-methylated DNA immunoprecipitation sequencing (5meDIP-seq) and 5-hydroxymethylated DNA immunoprecipitation sequencing (5hmeDIP-seq)) will be needed to address the relative contribution of these epigenetic modifications in specific DNA regions to aging.

Because evidence suggests that some epigenetic processes are cell type—specific, there is a great need for the study of epigenetic mechanisms that drive aging in different cell types and tissues to have a more complete picture of aging at the organismal level. The development of a more comprehensive epigenetic atlas of aging in other cell types and tissues, such as the one reported by Cheung and collaborators in immune cells,[78] would broaden the current understanding of epigenetic changes that take place during human aging, and help to identify universal modifications linked to aging that could be used as biomarkers of aging regardless of tissue or cell type.

Given the reversible nature of epigenetic modifications through modulation of the enzymes that deposit (writers) and remove (erasers) them, a deeper knowledge of the aging process will help us to identify potential therapeutical targets for delaying aging and prolonging human health span through epigenome fine-tuning. Modulation of the epigenetic regulators has been demonstrated to extend or reduce the life span in several model organisms, such as flies and yeast,[74,147–149] opening new and powerful perspectives for therapeutic interventions to improve health at the latest stages of life.

Epigenetics have been shown to be involved in the progression of aging, but given the complex nature of epigenetic regulation, more studies are needed to understand the wide range of these potentially reversible modifications that occur during human aging, and how they cross-talk with each other, contributing to the loss of transcriptional fidelity observed with the passage of time.

References

1. Trojer P, Reinberg D. Facultative heterochromatin: is there a distinctive molecular signature? *Mol Cell.* 2007;28(1):1–13. https://doi.org/10.1016/j.molcel.2007.09.011.
2. Saksouk N, Simboeck E, Déjardin J. Constitutive heterochromatin formation and transcription in mammals. *Epigenet Chromatin.* 2015;8(1):3. https://doi.org/10.1186/1756-8935-8-3.
3. Shah SG, Mandloi T, Kunte P, et al. HISTome2: a database of histone proteins, modifiers for multiple organisms and epidrugs. *Epigenet Chromatin.* 2020;13(1). https://doi.org/10.1186/s13072-020-00354-8.
4. Celeste A, Petersen S, Romanienko PJ, et al. Genomic instability in mice lacking histone H2AX. *Science.* 2002;296(5569):922–927. https://doi.org/10.1126/science.1069398.
5. Elsässer SJ, Noh K-M, Diaz N, Allis CD, Banaszynski LA. Histone H3.3 is required for endogenous retroviral element silencing in embryonic stem cells. *Nature.* 2015;522(7555):240–244. https://doi.org/10.1038/nature14345.
6. Creppe C, Janich P, Cantariño N, et al. MacroH2A1 regulates the balance between self-renewal and differentiation commitment in embryonic and adult stem cells. *Mol Cell Biol.* 2012;32(8):1442–1452. https://doi.org/10.1128/MCB.06323-11.
7. Creyghton MP, Markoulaki S, Levine SS, et al. H2AZ is enriched at polycomb complex target genes in ES cells and is necessary for lineage commitment. *Cell.* 2008;135(4):649–661. https://doi.org/10.1016/j.cell.2008.09.056.
8. Banaszynski LA, Wen D, Dewell S, et al. Hira-dependent histone H3.3 deposition facilitates PRC2 recruitment at developmental loci in ES cells. *Cell.* 2013;155(1):107–120. https://doi.org/10.1016/j.cell.2013.08.061.
9. Cheloufi S, Elling U, Hopfgartner B, et al. The histone chaperone CAF-1 safeguards somatic cell identity. *Nature.* 2015;528(7581):218–224. https://doi.org/10.1038/nature15749.
10. Yang HD, Kim P-J, Eun JW, et al. Oncogenic potential of histone-variant H2A.Z.1 and its regulatory role in cell cycle and epithelial-mesenchymal transition in liver cancer. *Oncotarget.* 2016;7(10):11412–11423. https://doi.org/10.18632/oncotarget.7194.
11. Kim J, Sturgill D, Sebastian R, et al. Replication stress shapes a protective chromatin environment across fragile genomic regions. *Mol Cell.* 2018;69(1):36–47.e7. https://doi.org/10.1016/j.molcel.2017.11.021.
12. Kovatcheva M, Liao W, Klein ME, et al. ATRX is a regulator of therapy induced senescence in human cells. *Nat Commun.* 2017;8(1):386. https://doi.org/10.1038/s41467-017-00540-5.
13. Gévry N, Chan HM, Laflamme L, Livingston DM, Gaudreau L. p21 transcription is regulated by differential localization of histone H2A. Z. *Genes Dev.* 2007;21(15):1869–1881. https://doi.org/10.1101/gad.1545707.

14. Duarte LF, Young ARJ, Wang Z, et al. Histone H3.3 and its proteolytically processed form drive a cellular senescence programme. *Nat Commun.* 2014;5(1). https://doi.org/10.1038/ncomms6210.

15. Rai TS, Cole JJ, Nelson DM, et al. HIRA orchestrates a dynamic chromatin landscape in senescence and is required for suppression of neoplasia. *Genes Dev.* 2014;28(24). https://doi.org/10.1101/gad.247528.114.

16. Singh N, Basnet H, Wiltshire TD, et al. Dual recognition of phosphoserine and phosphotyrosine in histone variant H2A.X by DNA damage response protein MCPH1. *Proc Natl Acad Sci U S A.* 2012;109(36):14381−14386. https://doi.org/10.1073/pnas.1212366109.

17. Lachat C, Boyer-Guittaut M, Peixoto P, Hervouet E. Epigenetic regulation of EMT (epithelial to mesenchymal transition) and tumor aggressiveness: a view on paradoxical roles of KDM6B and EZH2. *Epigenomes.* 2018;3(1). https://doi.org/10.3390/epigenomes3010001.

18. Schotta G, Lachner M, Sarma K, et al. A silencing pathway to induce H3-K9 and H4-K20 trimethylation at constitutive heterochromatin. *Genes Dev.* 2004;18(11):1251−1262. https://doi.org/10.1101/gad.300704.

19. Morgan MAJ, Shilatifard A. Reevaluating the roles of histone-modifying enzymes and their associated chromatin modifications in transcriptional regulation. *Nat Genet.* 2020;52(12). https://doi.org/10.1038/s41588-020-00736-4.

20. Taylor GCA, Eskeland R, Hekimoglu-Balkan B, Pradeepa MM, Bickmore WA. H4K16 acetylation marks active genes and enhancers of embryonic stem cells, but does not alter chromatin compaction. *Genome Res.* 2013;23(12):2053−2065. https://doi.org/10.1101/gr.155028.113.

21. Allis CD, Jenuwein T. The molecular hallmarks of epigenetic control. *Nat Rev Genet.* 2016. https://doi.org/10.1038/nrg.2016.59.

22. Hyun K, Jeon J, Park K, Kim J. Writing, erasing and reading histone lysine methylations. *Exp Mol Med.* 2017;49(4). https://doi.org/10.1038/emm.2017.11. e324−e324.

23. Bernstein BE, Mikkelsen TS, Xie X, et al. A bivalent chromatin structure marks key developmental genes in embryonic stem cells. *Cell.* 2006;125(2):315−326. https://doi.org/10.1016/j.cell.2006.02.041.

24. Azuara V, Perry P, Sauer S, et al. Chromatin signatures of pluripotent cell lines. *Nat Cell Biol.* 2006;8(5):532−538. https://doi.org/10.1038/ncb1403.

25. Bestor TH, Ingram VM. Two DNA methyltransferases from murine erythroleukemia cells: purification, sequence specificity, and mode of interaction with DNA. *Proc Natl Acad Sci U S A.* 1983;80(18). https://doi.org/10.1073/pnas.80.18.5559.

26. Okano M, Bell DW, Haber DA, Li E. DNA methyltransferases Dnmt3a and Dnmt3b are essential for de novo methylation and mammalian development. *Cell.* 1999;99(3). https://doi.org/10.1016/S0092-8674(00)81656-6.

27. Tahiliani M, Koh KP, Shen Y, et al. Conversion of 5-methylcytosine to 5-hydroxymethylcytosine in mammalian DNA by MLL partner TET1. *Science.* 2009;324(5929):930−935. https://doi.org/10.1126/science.1170116.

28. Ko M, Huang Y, Jankowska AM, et al. Impaired hydroxylation of 5-methylcytosine in myeloid cancers with mutant TET2. *Nature.* 2010;468(7325):839−843. https://doi.org/10.1038/nature09586.

29. He Y-F, Li B-Z, Li Z, et al. Tet-mediated formation of 5-carboxylcytosine and its excision by TDG in mammalian DNA. *Science.* 2011;80(6047):333. https://doi.org/10.1126/science.1210944.

30. Ito S, Shen L, Dai Q, et al. Tet proteins can convert 5-methylcytosine to 5-formylcytosine and 5-carboxylcytosine. *Science.* 2011;333(6047):1300−1303. https://doi.org/10.1126/science.1210597.

31. Wossidlo M, Nakamura T, Lepikhov K, et al. 5-Hydroxymethylcytosine in the mammalian zygote is linked with epigenetic reprogramming. *Nat Commun.* 2011;2(1):241. https://doi.org/10.1038/ncomms1240.
32. Ruzov A, Tsenkina Y, Serio A, et al. Lineage-specific distribution of high levels of genomic 5-hydroxymethylcytosine in mammalian development. *Cell Res.* 2011;21(9):1332−1342. https://doi.org/10.1038/cr.2011.113.
33. Nestor CE, Ottaviano R, Reddington J, et al. Tissue type is a major modifier of the 5-hydroxymethylcytosine content of human genes. *Genome Res.* 2012;22(3):467−477. https://doi.org/10.1101/gr.126417.111.
34. Cui X-L, Nie J, Ku J, et al. A human tissue map of 5-hydroxymethylcytosines exhibits tissue specificity through gene and enhancer modulation. *Nat Commun.* 2020;11(1):6161. https://doi.org/10.1038/s41467-020-20001-w.
35. Pastor WA, Pape UJ, Huang Y, et al. Genome-wide mapping of 5-hydroxymethylcytosine in embryonic stem cells. *Nature.* 2011;473(7347):394−397. https://doi.org/10.1038/nature10102.
36. The ENCODE Project Consortium. Identification and analysis of functional elements in 1% of the human genome by the ENCODE pilot project. *Nature.* 2007;447(7146). https://doi.org/10.1038/nature05874.
37. Lorenzi L, Chiu H-S, Avila Cobos F, et al. Publisher Correction: the RNA Atlas expands the catalog of human non-coding RNAs. *Nat Biotechnol.* 2021;39(11). https://doi.org/10.1038/s41587-021-00996-3, 1467-1467.
38. Hombach S, Kretz M. Non-coding RNAs: classification, biology and functioning. *Adv Exp Med Biol.* 2016. https://doi.org/10.1007/978-3-319-42059-2_1.
39. Batista PJ, Chang HY. Long noncoding RNAs: cellular address codes in development and disease. *Cell.* 2013;152(6):1298−1307. https://doi.org/10.1016/j.cell.2013.02.012.
40. Holoch D, Moazed D. RNA-mediated epigenetic regulation of gene expression. *Nat Rev Genet.* 2015;16(2):71−84. https://doi.org/10.1038/nrg3863.
41. Neve B, Jonckheere N, Vincent A, Van Seuningen I. Long non-coding RNAs: the tentacles of chromatin remodeler complexes. *Cell Mol Life Sci.* 2021;78(4):1139−1161. https://doi.org/10.1007/s00018-020-03646-0.
42. Kornienko AE, Guenzl PM, Barlow DP, Pauler FM. Gene regulation by the act of long non-coding RNA transcription. *BMC Biol.* 2013;11(1):59. https://doi.org/10.1186/1741-7007-11-59.
43. Penny GD, Kay GF, Sheardown SA, Rastan S, Brockdorff N. Requirement for Xist in X chromosome inactivation. *Nature.* 1996;379(6561):131−137. https://doi.org/10.1038/379131a0.
44. Galupa R, Heard E. X-chromosome inactivation: a crossroads between chromosome architecture and gene regulation. *Annu Rev Genet.* 2018;52(1):535−566. https://doi.org/10.1146/annurev-genet-120116-024611.
45. Ashe A, Sapetschnig A, Weick E-M, et al. piRNAs can trigger a multigenerational epigenetic memory in the germline of *C. elegans. Cell.* 2012;150(1):88−99. https://doi.org/10.1016/j.cell.2012.06.018.
46. Gu SG, Pak J, Guang S, Maniar JM, Kennedy S, Fire A. Amplification of siRNA in *Caenorhabditis elegans* generates a transgenerational sequence-targeted histone H3 lysine 9 methylation footprint. *Nat Genet.* 2012;44(2):157−164. https://doi.org/10.1038/ng.1039.
47. Shirayama M, Seth M, Lee H-C, et al. piRNAs initiate an epigenetic memory of nonself RNA in the *C. elegans* germline. *Cell.* 2012;150(1):65−77. https://doi.org/10.1016/j.cell.2012.06.015.
48. Liu J, Dou X, Chen C, et al. N 6 -methyladenosine of chromosome-associated regulatory RNA regulates chromatin state and transcription. *Science.* 2020;367(6477):580−586. https://doi.org/10.1126/science.aay6018.

49. Coppede F. Mutations involved in premature-ageing syndromes. *Appl Clin Genet.* 2021;14:279—295. https://doi.org/10.2147/TACG.S273525.

50. Kudlow BA, Kennedy BK, Monnat RJ. Werner and Hutchinson—Gilford progeria syndromes: mechanistic basis of human progeroid diseases. *Nat Rev Mol Cell Biol.* 2007;8(5). https://doi.org/10.1038/nrm2161.

51. De Sandre-Giovannoli A, Bernard R, Cau P, et al. Lamin A truncation in hutchinson-gilford progeria. *Science.* 2003;300(5628). https://doi.org/10.1126/science.1084125, 2055-2055.

52. Eriksson M, Brown WT, Gordon LB, et al. Recurrent de novo point mutations in lamin A cause Hutchinson—Gilford progeria syndrome. *Nature.* 2003;423(6937):293—298. https://doi.org/10.1038/nature01629.

53. Yu C-E, Oshima J, Fu Y-H, et al. Positional cloning of the werner's syndrome gene. *Science.* 1996;272(5259):258—262. https://doi.org/10.1126/science.272.5259.258.

54. Salzer MC, Lafzi A, Berenguer-Llergo A, et al. Identity noise and adipogenic traits characterize dermal fibroblast aging. *Cell.* 2018;175(6):1575—1590.e22. https://doi.org/10.1016/j.cell.2018.10.012.

55. de Magalhães JP, Passos JF. Stress, cell senescence and organismal ageing. *Mech Ageing Dev.* 2018;170:2—9. https://doi.org/10.1016/j.mad.2017.07.001.

56. López-otín C, Blasco MA, Partridge L, Serrano M, Kroemer G. The hallmarks of aging longevity. *Cell.* 2013;153(6):1194—1217. https://doi.org/10.1016/j.cell.2013.05.039.

57. Scaffidi P, Misteli T. Reversal of the cellular phenotype in the premature aging disease Hutchinson-Gilford progeria syndrome. *Nat Med.* 2005;11(4):440—445. https://doi.org/10.1038/nm1204.

58. Scaffidi P, Misteli T. Lamin A-dependent nuclear defects in human aging. *Science.* 2006;(5776):312. https://doi.org/10.1126/science.1127168.

59. Miller JD, Ganat YM, Kishinevsky S, et al. Human iPSC-based modeling of late-onset disease via progerin-induced aging. *Cell Stem Cell.* 2013;13(6):691—705. https://doi.org/10.1016/j.stem.2013.11.006.

60. Oberdoerffer P, Sinclair DA. The role of nuclear architecture in genomic instability and ageing. *Nat Rev Mol Cell Biol.* 2007;8(9):692—702. https://doi.org/10.1038/nrm2238.

61. Zhang W, Li J, Suzuki K, et al. A Werner syndrome stem cell model unveils heterochromatin alterations as a driver of human aging. *Science.* 2015;348(6239):1160—1163. https://doi.org/10.1126/science.aaa1356.

62. Zhang R, Poustovoitov MV, Ye X, et al. Formation of MacroH2A-containing senescence-associated heterochromatin foci and senescence driven by ASF1a and HIRA. *Dev Cell.* 2005;8(1). https://doi.org/10.1016/j.devcel.2004.10.019.

63. Narita M, Nuñ Ez S, Heard E, et al. *Rb-Mediated Heterochromatin Formation and Silencing of E2F Target Genes during Cellular Senescence.* Vol. 113. 2003.

64. Chandra T, Kirschner K, Thuret J-Y, et al. Independence of repressive histone marks and chromatin compaction during senescent heterochromatic layer formation. *Mol Cell.* 2012;47(2). https://doi.org/10.1016/j.molcel.2012.06.010.

65. Shah PP, Donahue G, Otte GL, et al. Lamin B1 depletion in senescent cells triggers large-scale changes in gene expression and the chromatin landscape. *Genes Dev.* 2013;27(16):1787—1799. https://doi.org/10.1101/gad.223834.113.

66. Zhang X, Liu X, Du Z, et al. The loss of heterochromatin is associated with multiscale three-dimensional genome reorganization and aberrant transcription during cellular senescence. *Genome Res.* 2021;31(7):1121—1135. https://doi.org/10.1101/gr.275235.121.

67. Henikoff S. Nucleosome destabilization in the epigenetic regulation of gene expression. *Nat Rev Genet.* 2008;9(1):15—26. https://doi.org/10.1038/nrg2206.

68. Moskowitz DM, Zhang DW, Hu B, et al. Epigenomics of human CD8 T cell differentiation and aging. *Sci Immunol.* 2017;2(8). https://doi.org/10.1126/sciimmunol.aag0192.

69. Ucar D, Márquez EJ, Chung CH, et al. The chromatin accessibility signature of human immune aging stems from CD8+ T cells. *J Exp Med.* 2017;214(10):3123–3144. https://doi.org/10.1084/jem.20170416.
70. Clapier CR, Cairns BR. The biology of chromatin remodeling complexes. *Annu Rev Biochem.* 2009;78(1). https://doi.org/10.1146/annurev.biochem.77.062706.153223.
71. Bartholomew B. Regulating the chromatin landscape: structural and mechanistic perspectives. *Annu Rev Biochem.* 2014;83(1). https://doi.org/10.1146/annurev-biochem-051810-093157.
72. Pegoraro G, Kubben N, Wickert U, Göhler H, Hoffmann K, Misteli T. Ageing-related chromatin defects through loss of the NURD complex. *Nat Cell Biol.* 2009;11(10):1261–1267. https://doi.org/10.1038/ncb1971.
73. Riedel CG, Dowen RH, Lourenco GF, et al. DAF-16 employs the chromatin remodeller SWI/SNF to promote stress resistance and longevity. *Nat Cell Biol.* 2013;15(5). https://doi.org/10.1038/ncb2720.
74. Ni Z, Ebata A, Alipanahiramandi E, Lee SS. Two SET domain containing genes link epigenetic changes and aging in *Caenorhabditis elegans. Aging Cell.* 2012;11(2):315–325. https://doi.org/10.1111/j.1474-9726.2011.00785.x.
75. Larson K, Yan SJ, Tsurumi A, et al. Heterochromatin formation promotes longevity and represses ribosomal RNA synthesis. *PLoS Genet.* 2012;8(1). https://doi.org/10.1371/journal.pgen.1002473.
76. Feser J, Truong D, Das C, et al. Elevated histone expression promotes life span extension. *Mol Cell.* 2010;39(5):724–735. https://doi.org/10.1016/j.molcel.2010.08.015.
77. O'Sullivan RJ, Kubicek S, Schreiber SL, Karlseder J. Reduced histone biosynthesis and chromatin changes arising from a damage signal at telomeres. *Nat Struct Mol Biol.* 2010;17(10):1218–1225. https://doi.org/10.1038/nsmb.1897.
78. Cheung P, Vallania F, Warsinske HC, et al. Single-cell chromatin modification profiling reveals increased epigenetic variations with aging. *Cell.* 2018;173(6):1385–1397.e14. https://doi.org/10.1016/j.cell.2018.03.079.
79. Ivanov A, Pawlikowski J, Manoharan I, et al. Lysosome-mediated processing of chromatin in senescence. *J Cell Biol.* 2013;202(1):129–143. https://doi.org/10.1083/jcb.201212110.
80. Talbert PB, Henikoff S. Histone variants at a glance. *J Cell Sci.* 2021;134(6). https://doi.org/10.1242/jcs.244749.
81. Rogakou EP, Pilch DR, Orr AH, Ivanova VS, Bonner WM. DNA double-stranded breaks induce histone H2AX phosphorylation on serine 139. *J Biol Chem.* 1998;273(10). https://doi.org/10.1074/jbc.273.10.5858.
82. Podhorecka M, Skladanowski A, Bozko P. H2AX phosphorylation: its role in DNA damage response and cancer therapy. *J Nucleic Acids.* 2010;2010. https://doi.org/10.4061/2010/920161.
83. Rogakou EP, Sekeri–Pataryas KE. Histone variants of H2A and H3 families are regulated during in vitro aging in the same manner as during differentiation. *Exp Gerontol.* 1999;34(6). https://doi.org/10.1016/S0531-5565(99)00046-7.
84. Sedelnikova OA, Horikawa I, Zimonjic DB, Popescu NC, Bonner WM, Barrett JC. Senescing human cells and ageing mice accumulate DNA lesions with unrepairable double-strand breaks. *Nat Cell Biol.* 2004;6(2). https://doi.org/10.1038/ncb1095.
85. Contrepois K, Coudereau C, Benayoun BA, et al. Histone variant H2A.J accumulates in senescent cells and promotes inflammatory gene expression. *Nat Commun.* 2017;8. https://doi.org/10.1038/ncomms14995.
86. Rübe CE, Bäumert C, Schuler N, et al. Human skin aging is associated with increased expression of the histone variant H2A.J in the epidermis. *Npj Aging Mech Dis.* 2021;7(1). https://doi.org/10.1038/s41514-021-00060-z.

87. Pehrson JR, Fried VA. MacroH2A, a core histone containing a large nonhistone region. *Science*. 1992;257(5075):1398—1400. https://doi.org/10.1126/science.1529340.
88. Lavigne MD, Vatsellas G, Polyzos A, et al. Composite macroH2A/NRF-1 nucleosomes suppress noise and generate robustness in gene expression. *Cell Rep*. 2015;11(7):1090—1101. https://doi.org/10.1016/j.celrep.2015.04.022.
89. Ni K, Ren J, Xu X, et al. LSH mediates gene repression through macroH2A deposition. *Nat Commun*. 2020;11(1):5647. https://doi.org/10.1038/s41467-020-19159-0.
90. Douet J, Corujo D, Malinverni R, et al. MacroH2A histone variants maintain nuclear organization and heterochromatin architecture. *J Cell Sci*. *January*. 2017. https://doi.org/10.1242/jcs.199216.
91. Maze I, Wenderski W, Noh K-M, et al. Critical role of histone turnover in neuronal transcription and plasticity. *Neuron*. 2015;87(1). https://doi.org/10.1016/j.neuron.2015.06.014.
92. Szenker E, Ray-Gallet D, Almouzni G. The double face of the histone variant H3.3. *Cell Res*. 2011;21(3):421—434. https://doi.org/10.1038/cr.2011.14.
93. Giunta S, Hervé S, White RR, et al. CENP-A chromatin prevents replication stress at centromeres to avoid structural aneuploidy. *Proc Natl Acad Sci U S A*. 2021;118(10). https://doi.org/10.1073/pnas.2015634118.
94. Maehara K, Takahashi K, Saitoh S. CENP-A reduction induces a p53-dependent cellular senescence response to protect cells from executing defective mitoses. *Mol Cell Biol*. 2010;30(9). https://doi.org/10.1128/MCB.01318-09.
95. Lee S-H, Itkin-Ansari P, Levine F. CENP-A, a protein required for chromosome segregation in mitosis, declines with age in islet but not exocrine cells. *Aging (Albany NY)*. 2010;2(11). https://doi.org/10.18632/aging.100220.
96. Hoffmann S, Izquierdo HM, Gamba R, et al. A genetic memory initiates the epigenetic loop necessary to preserve centromere position. *EMBO J*. 2020;39(20). https://doi.org/10.15252/embj.2020105505.
97. Hashimoto H, Vertino PM, Cheng X. Molecular coupling of DNA methylation and histone methylation. *Epigenomics*. 2010;2(5). https://doi.org/10.2217/epi.10.44.
98. Uckelmann M, Sixma TK. Histone ubiquitination in the DNA damage response. *DNA Repair*. 2017;56. https://doi.org/10.1016/j.dnarep.2017.06.011.
99. Martinez-Jimenez CP, Eling N, Chen H-C, et al. Aging increases cell-to-cell transcriptional variability upon immune stimulation. *Science*. 2017;355(6332):1433—1436. https://doi.org/10.1126/science.aah4115.
100. Enge M, Arda HE, Mignardi M, et al. Single-cell analysis of human pancreas reveals transcriptional signatures of aging and somatic mutation patterns. *Cell*. 2017;171(2):321—330.e14. https://doi.org/10.1016/j.cell.2017.09.004.
101. Shumaker DK, Dechat T, Kohlmaier A, et al. *Mutant Nuclear Lamin A Leads to Progressive Alterations of Epigenetic Control in Premature Aging*; 2006. www.pnas.orgcgidoi10.1073pnas.0602569103.
102. McCord RP, Nazario-Toole A, Zhang H, et al. Correlated alterations in genome organization, histone methylation, and DNA-lamin A/C interactions in Hutchinson-Gilford progeria syndrome. *Genome Res*. 2013;23(2):260—269. https://doi.org/10.1101/gr.138032.112.
103. Tsurumi A, Li W. Global heterochromatin loss. *Epigenetics*. 2012;7(7). https://doi.org/10.4161/epi.20540.
104. Villeponteau B. The heterochromatin loss model of aging. *Exp Gerontol*. 1997;32(4—5). https://doi.org/10.1016/S0531-5565(96)00155-6.
105. Eisenberg T, Knauer H, Schauer A, et al. Induction of autophagy by spermidine promotes longevity. *Nat Cell Biol*. 2009;11(11):1305—1314. https://doi.org/10.1038/ncb1975.
106. Sasaki T, Maier B, Bartke A, Scrable H. Progressive loss of SIRT1 with cell cycle withdrawal. *Aging Cell*. 2006;5(5). https://doi.org/10.1111/j.1474-9726.2006.00235.x.

107. Greer EL, Maures TJ, Hauswirth AG, et al. Members of the H3K4 trimethylation complex regulate lifespan in a germline-dependent manner in *C. elegans*. *Nature*. 2010;466(7304). https://doi.org/10.1038/nature09195.

108. Jung M, Pfeifer GP. Aging and DNA methylation. *BMC Biol*. 2015;13(1):7. https://doi.org/10.1186/s12915-015-0118-4.

109. Day K, Waite LL, Thalacker-Mercer A, et al. Differential DNA methylation with age displays both common and dynamic features across human tissues that are influenced by CpG landscape. *Genome Biol*. 2013;14(9). https://doi.org/10.1186/gb-2013-14-9-r102.

110. Teschendorff AE, Menon U, Gentry-Maharaj A, et al. Age-dependent DNA methylation of genes that are suppressed in stem cells is a hallmark of cancer. *Genome Res*. 2010;20(4):440−446. https://doi.org/10.1101/gr.103606.109.

111. Bird AP, Wolffe AP. Methylation-induced repression— belts, braces, and chromatin. *Cell*. 1999;99(5):451−454. https://doi.org/10.1016/S0092-8674(00)81532-9.

112. Lorincz MC, Schubeler D, Hutchinson SR, Dickerson DR, Groudine M. DNA methylation density influences the stability of an epigenetic imprint and dnmt3a/b-independent de novo methylation. *Mol Cell Biol*. 2002;22(21):7572−7580. https://doi.org/10.1128/MCB.22.21.7572-7580.2002.

113. Wilson VL, Jones PA. DNA methylation decreases in aging but not in immortal cells. *Science*. 1983;(4601):220. https://doi.org/10.1126/science.6844925.

114. Buscarlet M, Tessier A, Provost S, Mollica L, Busque L. Human blood cell levels of 5-hydroxymethylcytosine (5hmC) decline with age, partly related to acquired mutations in TET2. *Exp Hematol*. 2016;44(11):1072−1084. https://doi.org/10.1016/j.exphem.2016.07.009.

115. Reynolds LM, Taylor JR, Ding J, et al. Age-related variations in the methylome associated with gene expression in human monocytes and T cells. *Nat Commun*. 2014;5(1):5366. https://doi.org/10.1038/ncomms6366.

116. Shchukina I, Bagaitkar J, Shpynov O, et al. Enhanced epigenetic profiling of classical human monocytes reveals a specific signature of healthy aging in the DNA methylome. *Nat Aging*. 2021;1(1):124−141. https://doi.org/10.1038/s43587-020-00002-6.

117. Koch CM, Wagner W. Epigenetic-aging-signature to determine age in different tissues. *Aging (Albany NY)*. 2011;3(10):1018−1027. https://doi.org/10.18632/aging.100395.

118. Bocklandt S, Lin W, Sehl ME, et al. Epigenetic predictor of age. *PLoS One*. 2011;6(6). https://doi.org/10.1371/journal.pone.0014821.

119. Horvath S. *DNA Methylation Age of Human Tissues and Cell Types*. Vol. 14. 2013. http://genomebiology.com//14/10/R115.

120. Hannum G, Guinney J, Zhao L, et al. Genome-wide methylation profiles reveal quantitative views of human aging rates. *Mol Cell*. 2013;49(2):359−367. https://doi.org/10.1016/j.molcel.2012.10.016.

121. Horvath S, Erhart W, Brosch M, et al. Obesity accelerates epigenetic aging of human liver. *Proc Natl Acad Sci U S A*. 2014;111(43). https://doi.org/10.1073/pnas.1412759111.

122. Simpson DJ, Chandra T. Epigenetic age prediction. *Aging Cell*. 2021;20(9):1−20. https://doi.org/10.1111/acel.13452.

123. Zhang Y, Wilson R, Heiss J, et al. DNA methylation signatures in peripheral blood strongly predict all-cause mortality. *Nat Commun*. 2017;8. https://doi.org/10.1038/ncomms14617.

124. Levine ME, Lu AT, Quach A, et al. *An Epigenetic Biomarker of Aging for Lifespan and Healthspan*. Vol. 10. 2018. www.aging-us.com.

125. Lu AT, Quach A, Wilson JG, et al. DNA methylation GrimAge strongly predicts lifespan and healthspan. *Aging (Albany NY)*. 2019;11(2):303−327. https://doi.org/10.18632/aging.101684.

126. Lu AT, Seeboth A, Tsai P-C, et al. DNA methylation-based estimator of telomere length. *Aging (Albany NY)*. 2019;11(16):5895–5923. https://doi.org/10.18632/aging.102173.

127. Belsky DW, Caspi A, Arseneault L, et al. Quantification of the pace of biological aging in humans through a blood test, the DunedinPoAm DNA methylation algorithm. *Elife*. 2020;9. https://doi.org/10.7554/eLife.54870.

128. Kerepesi C, Zhang B, Lee SG, Trapp A, Gladyshev VN. Epigenetic clocks reveal a rejuvenation event during embryogenesis followed by aging. *Sci Adv*. 2021;7(26):1–12. https://doi.org/10.1126/sciadv.abg6082.

129. Noren Hooten N, Abdelmohsen K, Gorospe M, Ejiogu N, Zonderman AB, Evans MK. microRNA expression patterns reveal differential expression of target genes with age. *PLoS One*. 2010;5(5). https://doi.org/10.1371/journal.pone.0010724.

130. Hofmann P, Sommer J, Theodorou K, et al. Long non-coding RNA H19 regulates endothelial cell aging via inhibition of STAT3 signalling. *Cardiovasc Res*. 2019;115(1). https://doi.org/10.1093/cvr/cvy206.

131. Cai D, Han J-DJ. Aging-associated lncRNAs are evolutionarily conserved and participate in NFκB signaling. *Nat Aging*. 2021;1(5). https://doi.org/10.1038/s43587-021-00056-0.

132. Takahashi K, Yamanaka S. Induction of pluripotent stem cells from mouse embryonic and adult fibroblast cultures by defined factors. *Cell*. 2006;126(4). https://doi.org/10.1016/j.cell.2006.07.024.

133. Takahashi K, Tanabe K, Ohnuki M, et al. Induction of pluripotent stem cells from adult human fibroblasts by defined factors. *Cell*. 2007;131(5):861–872. https://doi.org/10.1016/j.cell.2007.11.019.

134. Rando TA, Chang HY. Aging, rejuvenation, and epigenetic reprogramming: resetting the aging clock. *Cell*. 2012;148(1–2):46–57. https://doi.org/10.1016/j.cell.2012.01.003.

135. Zhang W, Qu J, Liu GH, Belmonte JCI. The ageing epigenome and its rejuvenation. *Nat Rev Mol Cell Biol*. 2020;21(3):137–150. https://doi.org/10.1038/s41580-019-0204-5.

136. Manukyan M, Singh PB. Epigenome rejuvenation: HP1β mobility as a measure of pluripotent and senescent chromatin ground states. *Sci Rep*. 2014;4(1):4789. https://doi.org/10.1038/srep04789.

137. Ocampo A, Reddy P, Martinez-Redondo P, et al. In vivo amelioration of age-associated hallmarks by partial reprogramming. *Cell*. 2016;167(7):1719–1733. https://doi.org/10.1016/j.cell.2016.11.052.

138. Olova N, Simpson DJ, Marioni RE, Chandra T. Partial reprogramming induces a steady decline in epigenetic age before loss of somatic identity. *Aging Cell*. 2019;18(1). https://doi.org/10.1111/acel.12877.

139. Sarkar TJ, Quarta M, Mukherjee S, et al. Transient non-integrative expression of nuclear reprogramming factors promotes multifaceted amelioration of aging in human cells. *Nat Commun*. 2020;11(1). https://doi.org/10.1038/s41467-020-15174-3.

140. Gill D, Parry A, Santos F, et al. Multi-omic rejuvenation of human cells by maturation phase transient reprogramming. *Elife*. 2022;11. https://doi.org/10.7554/eLife.71624.

141. Lu Y, Brommer B, Tian X, et al. Reprogramming to recover youthful epigenetic information and restore vision. *Nature*. 2020;588(7836):124–129. https://doi.org/10.1038/s41586-020-2975-4.

142. Fahy GM, Brooke RT, Watson JP, et al. Reversal of epigenetic aging and immunosenescent trends in humans. *Aging Cell*. 2019;18(6). https://doi.org/10.1111/acel.13028.

143. De Lucia F, Dean C. Long non-coding RNAs and chromatin regulation. *Curr Opin Plant Biol*. 2011;14(2):168–173. https://doi.org/10.1016/j.pbi.2010.11.006.

144. Long Y, Hwang T, Gooding AR, Goodrich KJ, Rinn JL, Cech TR. RNA is essential for PRC2 chromatin occupancy and function in human pluripotent stem cells. *Nat Genet*. 2020;52(9):931−938. https://doi.org/10.1038/s41588-020-0662-x.
145. Guallar D, Bi X, Pardavila JA, et al. RNA-dependent chromatin targeting of TET2 for endogenous retrovirus control in pluripotent stem cells. *Nat Genet*. 2018;50(3):443−451. https://doi.org/10.1038/s41588-018-0060-9.
146. Xu W, Li J, He C, et al. METTL3 regulates heterochromatin in mouse embryonic stem cells. *Nature*. 2021;591(7849):317−321. https://doi.org/10.1038/s41586-021-03210-1.
147. Siebold AP, Banerjee R, Tie F, Kiss DL, Moskowitz J, Harte PJ. Polycomb Repressive Complex 2 and Trithorax modulate Drosophila longevity and stress resistance. *Proc Natl Acad Sci U S A*. 2010;107(1). https://doi.org/10.1073/pnas.0907739107.
148. Maures TJ, Greer EL, Hauswirth AG, Brunet A. The H3K27 demethylase UTX-1 regulates *C. elegans* lifespan in a germline-independent, insulin-dependent manner. *Aging Cell*. 2011;10(6). https://doi.org/10.1111/j.1474-9726.2011.00738.x.
149. Li L, Greer C, Eisenman RN, Secombe J. Essential functions of the histone demethylase lid. *PLoS Genet*. 2010;6(11). https://doi.org/10.1371/journal.pgen.1001221.
150. Ferrand J, Rondinelli B, Polo SE. Histone variants: guardians of genome integrity. *Cells*. 2020;9(11). https://doi.org/10.3390/cells9112424.
151. Tessarz P, Kouzarides T. Histone core modifications regulating nucleosome structure and dynamics. *Nat Rev Mol Cell Biol*. 2014;15(11):703−708. https://doi.org/10.1038/nrm3890.
152. Duempelmann L, Skribbe M, Bühler M. Small RNAs in the transgenerational inheritance of epigenetic information. *Trends Genet*. 2020;36(3):203−214. https://doi.org/10.1016/j.tig.2019.12.001.
153. Oshima J, Sidorova JM, Monnat RJ. Werner syndrome: clinical features, pathogenesis and potential therapeutic interventions. *Ageing Res Rev*. 2017;33:105−114. https://doi.org/10.1016/j.arr.2016.03.002.
154. Di Micco R, Krizhanovsky V, Baker D, d'Adda di Fagagna F. Cellular senescence in ageing: from mechanisms to therapeutic opportunities. *Nat Rev Mol Cell Biol*. 2021;22(2):75−95. https://doi.org/10.1038/s41580-020-00314-w.
155. Brunet A. Old and new models for the study of human ageing. *Nat Rev Mol Cell Biol*. 2020;21(9):491−493. https://doi.org/10.1038/s41580-020-0266-4.
156. Sen P, Shah PP, Nativio R, Berger SL. Epigenetic mechanisms of longevity and aging. *Cell*. 2016;166(4):822−839. https://doi.org/10.1016/j.cell.2016.07.050.
157. Benayoun BA, Pollina EA, Brunet A. Epigenetic regulation of ageing: linking environmental inputs to genomic stability. *Nat Rev Mol Cell Biol*. 2015;16(10):593−610. https://doi.org/10.1038/nrm4048.

SECTION II

Metabolism, homeostasis, and communication

CHAPTER 3

Nutrient sensing and aging

Lili Yang

Department of Nutrition, School of Public Health, Sun Yat-sen University, Guangzhou, China

Under normal conditions, nutrient sensing signals sense the levels of certain nutrients in the fasted or fed state. Corresponding cell signaling pathways are activated or inactivated to ensure proper cellular functions. The nutrient sensing ability of cells allows them to recognize and respond to the current state of nutrient levels. Cells require a constant supply of nutrients; yet our nutrient intake is not always continuous. As a consequence, cells must be able to store extra nutrients during times of nutrient excess. During fasting, cells sense the deficiency of nutrients and trigger a series of signaling pathways, depending on the type and extent of nutrients that are in deficit. Cells constantly manage energy consumption levels according to the availability of nutrients to preserve the production of adenosine triphosphate (ATP) and support the overall survival and maintenance of normal functions. When cells use this stored energy for any manner of cellular processes, ATP is converted to adenosine diphosphate (ADP), and further to adenosine monophosphate (AMP). It is essential for nutrient sensing signals to respond and minimize energy consumption to avoid exhausting the reserved resources. Meanwhile, certain nutrient sensing signals are modulated to restore the cellular energy supply, such as increasing nutrient intake, activating related energy-producing pathways, and turning certain molecules into nutrients.

Current knowledge tells us that AMP-activated protein kinase (AMPK) signaling has a significant role in regulating energy metabolism and controlling cellular functions such as aging.[1,2] With aging, cellular ATP levels significantly decrease in both animals and humans. Unlike under normal conditions, an insufficiency of ATP or an increased ratio of AMP/ATP will activate AMPK and its downstream signals. However, the reduced capacity of AMPK to respond to stresses with aging can augment aging. It is known that AMPK regulates some signaling pathways of aging, including autophagy, endoplasmic reticulum (ER) stress, mitochondrial dysfunction, oxidative stress, inflammation, and DNA damage repair.[2–4] For instance, abundant studies have revealed that overexpression of AMPK promotes longevity in different models.[5–7] Besides the unregulated AMPK pathway,

Molecular, Cellular, and Metabolic Fundamentals of Human Aging
ISBN 978-0-323-91617-2
https://doi.org/10.1016/B978-0-323-91617-2.00001-8

© 2023 Elsevier Inc.
All rights reserved.

other nutrient sensing signals are modulated with aging. Decreased levels of cellular NAD with aging inactivate the sirtuin (SIRT) pathway. In addition, inactivation of the mammalian target of rapamycin (mTOR) pathway, a sensor of high levels of amino acids, usually can be seen with aging. The nutrient sensing signaling pathways are interconnected, such as AMPK and SIRT, AMPK and mTOR, and mTOR and SIRT. AMPK is at the center of nutrient sensing, because deficiency of any of the energy-yielding nutrients will eventually cause a shortage of ATP in cells. In this chapter, we discuss nutrient sensing signals in terms of AMPK, mTOR, SIRT, and insulin-like growth factor 1 (IGF-1). We also cover the use of calorie restriction (CR) on the corresponding signaling pathways to fight against aging.

1. AMPK signaling pathway

AMPK is a ubiquitously expressed serine/threonine protein kinase activated by low cellular nutrient availability status. Once activated, AMPK triggers catalytic reactions to generate ATP while inhibiting anabolic processes that consume ATP, resulting in the restoration of cellular energy homeostasis.

The activation of AMPK signaling predominantly occurs via three upstream kinases, including liver kinase B1 (LKB1), Ca^{2+}, calmodulin-dependent protein kinase kinase β, and transforming growth factor β–activated kinase 1.[8–14] Downstream mediators of AMPK signaling contribute to regulating the autophagy pathway and the metabolism that restores energy homeostasis by increasing mitochondrial biogenesis. Correspondingly, activation of AMPK leads to the phosphorylation of key metabolic substrates and transcriptional regulators including glucose transporter 1/4, peroxisome proliferator-activated receptor gamma coactivator-1α (PGC-1α), and forkhead box transcription factors (FOXOs).[15] Evidence in loss-of-function studies showed that AMPK knockout is able to induce failure in autophagy and mitophagy, downregulate essential cardiac signaling molecules, and induce age-related changes in cardiac function.[16] Muscle-specific AMPK knockout showed reduced muscle function, decreased mitochondrial function, and accumulation of autophagy/mitophagy proteins in an aged mice model.[17] Downregulated phosphorylation of AMPK and autophagy also mediates aging in the kidneys and renal tubular cells and in the aging brain[18,19]. AMPK activity is also required to complete mitosis, and the active form of AMPK is localized to the mitotic apparatus in dividing cells.

A study reported that AMPK activates glycolysis by phosphorylating PFKFB3 during mitotic arrest, which suggests a role for AMPK in preserving cellular energy homeostasis when the cell cycle is perturbed during cellular senescence.[20]

It seems evident that aging increases the function of the inhibitory mechanisms of AMPK signaling.[21] AMPK activation is linked to many pathways enhancing longevity. For example, it inhibits inflammation, suppresses IGF-1/insulin/mTOR signaling, stimulates SIRT signaling, and prevents mitochondrial disturbances.[4,6,22−24] This suggests that AMPK activation is a promising measure for managing aging.

2. mTOR pathway

mTOR is an evolutionarily conserved serine-threonine kinase that senses a variety of signals directing cellular and organismal responses. It is in the PI3K-related kinase family, which forms the catalytic subunit of two distinct protein complexes, known as mTOR Complex 1 (mTORC1) and 2 (mTORC2). mTORC1 regulates cell growth and metabolism, mTORC2 controls proliferation and survival.[25] Activation of mTORC1 requires the presence of sufficient nutrients and energy sources, such as amino acids, glucose, lipids, oxygen, and a high ATP/AMP ratio.[26,27] Inhibition of mTOR as well as genetic studies found that reduced mTOR signaling promotes longevity in *Drosophila*, yeast, and mice.[28−31] Inhibition of mTORC1 with rapamycin is the only known pharmacologic treatment that increases the life span in all model organisms studied.[30,32−34]

Although studies show that mTOR signaling has a key role in mammalian aging, the mechanism through which this occurs is still unclear. mTORC1 is a major regulator of autophagy, which helps clear damaged proteins and organelles such as mitochondria, the accumulation of which are also associated with aging and aging-related diseases.[23,28,35] The nutritional status of the cells, especially the amino acid levels, dictates the recruitment of mTORC1 to the lysosomal membrane and its subsequent activation. Active mTORC1 phosphorylates several targets, including transcription factors TFEB and FOXO. Phosphorylation of TFEB, or possibly of FOXO, occurs at the lysosomal membrane and leads to retention in the cytosol, inhibiting autophagy.[35,36] mTORC1 inhibits autophagy and positively regulates growth, messenger RNA translation, and ribosomal biogenesis, which are related possibilities for inducing aging.[37]

mTOR inhibition by rapamycin extends the life span and delays the onset of age-associated diseases in animal models, with similar longevity-enhancing effects observed in humans. However, prolonged rapamycin treatment in humans has the potential for side effects such as glucose intolerance and insulin resistance.[30,34] Detailed molecular mechanisms and further preclinical studies are needed before applying mTOR inhibitors against aging.

3. SIRT pathway

SIRT is a family of nicotinamide adenine dinucleotide (NAD) consuming proteins that have deacetylase activities catalyzing the removal of acetyl or acyl groups from lysins of SIRT substrate proteins accompanied by their transfer to ADP ribose. SIRT signaling pathways are nutrient sensors regulated by AMPK or mTOR, or directly regulated by nutrient deficiency induced by a lack of NAD[38]. NAD deficiency is a consequence of decreased energy metabolism, especially intermediate metabolism in the citric acid cycle in mitochondria.[22,23] The decline in NAD was first noticed in transgenic mice overexpressing SIRT1 in pancreatic β cells,[38] which lost the ability to experience glucose-stimulated insulin secretion when they aged. NAD levels have been shown to decline approximately twofold in old worms and in multiple tissues including liver and skeletal muscle in aged mice.[39,40] The systemic decline in NAD has emerged as a plausible explanation for why aging affects SIRTs.

Mammalian SIRT1−7 have different[41] enzymatic activities and subcellular compartmentations. SIRT1 and SIRT6 are closely related to aging and life span extension, which are found in the nucleus; some SIRT1 are found outside the nucleus in the cytoplasm. The levels of SIRT1 and SIRT6 are reported to decrease in multiple senescent cells of mouse fibroblasts, lung epithelial cells, and endothelial cells.[42−45] Aging-induced inactivation of SIRT1 has a direct effect on mitochondrial biogenesis, oxidative metabolism, and associated antioxidant defense pathways, leading to damage to the electron transport chain and a decline in mitochondrial function.[41,46] Failure of SIRT1 to deacetylate downstream effectors may lead to a reduction in mitochondrial antioxidant defenses.[47,48] The inhibition of SIRT1 and SIRT6 expression via pharmacologic inhibitors or short interfering RNAs promotes premature senescence-like phenotypes in endothelial cells.[49,50] Conversely, overexpression of SIRT1 and SIRT6 suppresses stress-induced cellular senescence in human coronary artery cells, porcine aortic endothelial cells, and lung cells.[51−53] Overexpressing

SIRT6 significantly increased the life span in wild-type mice,[54] whereas SIRT6-deficient mice had shorter life spans than those of controls.[55] SIRT1 and SIRT6 have antiaging roles in various organs. In the central nervous system, a drop in hippocampal SIRT1 level or activity was noted in the aged rat brain.[56] SIRT1 mediates central circadian control in mice. Young mice lacking brain SIRT1 experience aging-dependent circadian changes, whereas mice over-expressing SIRT1 are protected from aging, indicating that a decline in SIRT1 activity may have an important role in aging.[57,58] Vascular aging involves the senescence of endothelial and vascular smooth muscle cells.[41] Both replicative senescence with telomere attrition and stress-induced premature senescence induce vascular cell growth arrest and a loss of vascular homeostasis, and contribute to the initiation and progression of cardiovascular diseases.[59] In aging mice, SIRT1 expression decreased in endothelial cells or vascular smooth muscle cells. Accordingly, the deacetylation of p53 by SIRT1 decreased, which caused cells to lose the ability to control oxidative stress and induced premature senescence. SIRT1 also regulates the FOXO1 and PAI-1 pathway to prevent senescence in endothelial cells.[44,50,60]

Plenty of studies have revealed that increasing cellular NAD levels has an antiaging effect. The increased NAD supply induced by overexpression of nicotinamide phosphoribosyltransferase (NAMPT) protected vascular cells from oxidative stress-induced senescence.[61] Supplementation with nicotinamide mononucleotide (NMN) has been shown to restore NAD levels and prevent diet- and age-induced type 2 diabetes in mice.[40,62] NMN restored the aging-induced reduction of NAD in pancreatic β cells and reversed the compromised insulin secretion ability in SIRT1-overexpressing aged mice, or decrease mitochondrial damage, and reverse age-associated physiologic decline at the cellular and organismal levels.[46,63] Another NAD intermediate, nicotinamide riboside (NR), can be converted to NAD after conversion to NMN via NR kinase.[64,65] Like NMN, NR boosts NAD levels in worms and mice and counteracts aging. NR supplementation also increases mitochondrial NAD levels and stimulates SIRT-mediated deacetylation of cytosolic or mitochondrial proteins, which may also be related to the anti-aging effect of NAD.[66-70]

4. IGF-1 pathway

IGF-1 is produced mainly in the liver in response to growth hormone (GH) stimulation. As a consequence of the decline in GH synthesis and release, systemic IGF-1 levels decline with advancing age. Although insulin binds to insulin receptors whereas IGF-1 binds to IGF1R, they have high structural similarity and share almost identical intracellular signaling molecules such as Akt and FOXO. Akt is a kinase regulating growth and metabolism in various tissues. FOXO is a key transcription factor regulating the expression of genes including antioxidative enzymes. In humans, serum IGF-1 is the highest during adolescence and may decline prematurely in the life course starting in middle age; it parallels the onset of aging.[71]

The decrease in IGF-1 with aging is temporally associated with the appearance of cognitive impairment. Supplementation of IGF-1 has been shown to reverse this deficit in humans.[72,73] On a cellular basis, when IGF-1 is reduced, the structural complexity of neurons is decreased and long-term potentiation is impaired, whereas the exogenous application of IGF-1 induces increased neuron complexity within the hippocampus as well as enhanced excitability and long-term potentiation reverses the aging-induced deficit in cognition.[74,75] The age-related decline in circulating levels of IGF-1 has a key role in neurovascular dysfunction. The critical role of IGF-1 in cerebromicrovascular endothelial health and protection of cognitive health was shown via an endothelium-specific or astrocyte-specific knockout of IGF-1R in mouse models.[76-79] By contrast, mutations decreasing insulin/IGF-1 signaling dramatically increased the life span in *Caenorhabditis elegans* and *Drosophila melanogaster* and in several mouse models.[80] In *C. elegans*, insulin signaling in response to food intake leads to the secretion of insulin-like peptides that bind to insulin/IGF-1–like receptor. Mutations in the receptor abolished the life span extensions.[81] Decreased GH/IGF-1 signaling has been shown to extend longevity in a wide variety of species including worms, fruit flies, yeast, and mice. Mice with a mutated allele of the IGF-1R are reported to be smaller than wild-type mice. These mice exhibit a marked reduction in IGF-1—induced intracellular signaling, which predisposes them to become insulin–resistant with age. This suggests that although decreased IGF-1 signaling might increase the life span, it might decrease the health span.

5. Calorie restriction and aging

CR is a type of nutritional intervention characterized by reduced dietary intake below energy requirements. Maintaining optimal nutrition CR requires that while energy intake is restricted, the diet should provide sufficient energy for metabolic homeostasis and be rich in micronutrients and dietary fiber. The optimal dose of CR has a key role in preventing and treating obesity and its complications, as well as antiaging effects. Data in nonhuman primates indicate that CR slows age-related muscle loss, hearing loss, and brain degeneration, and extends the health span and life span in rodent and primate models.[82,83] In humans undergoing 10%–30% CR, declines in various markers of oxidative stress (DNA damage and superoxide dismutase (SOD) activity)[84–86] and biomarkers of longevity (fasting insulin level and body temperature) were observed.[87,88]

Preclinical studies showed that dietary restriction can delay aging and promote longevity by modulating key nutrient-sensing pathways.[89] CR and activated signaling through AMPK can increase the life span in both mice and worms.[90] Through a cellular perspective, improvements in longevity seem to be mediated through a series of signaling pathways that improve the health span by lowering inflammation and oxidative stress.[91] Fasting initiates a rise in the AMP/ATP ratio and the activation of AMPK, leading to the inhibition of mTOR and an upregulation of autophagy.[92–94] A series of transcription factors such as FOXOs and PGC-1α are activated to trigger the expression of genes improving blood glucose regulation, mitochondrial biogenesis, and cell survival.[95,96]

However, long-term CR might result in health concerns. The potential lowering effect on IGF-1 levels of long-term CR has led to concern regarding its stimulation of muscle protein breakdown, a process that subsequently results in shrunken bone and muscle mass and a decrease muscle aerobic capacity in proportion to the reduction in body weight.[95,97,98] The literature indicates that in mouse studies, long-term severe CR may lead to compromised production of essential hormones. However, in human trials, long-term CR intervention without malnutrition had no effect on altering serum concentrations of IGF-1.[99] Evidence demonstrates that protein restriction can lower IGF-1 concentrations, indicating that protein restriction can potentially inhibit IGF-1 signaling and downregulate mTOR activity, leading to the activation of autophagy and inhibition of protein synthesis.[96,100,101] This agrees with the requirement that CR should supply sufficient nutrients for the body. A deficiency of certain nutrients may trigger certain nutrient sensing signaling pathways to induce side effects.

Moderate CR (10%–30% energy reduction) with sufficient protein and micronutrient intake is recommended for antiaging and to promote health. Concerns about long-term severe CR need to be considered by health professionals to prevent these side effects.

6. Conclusion

Nutrient sensing signals are interconnected in a complex network. During aging, AMPK, SIRT, and IGF-1 pathways are downregulated whereas the mTOR pathway is upregulated. Based on evidence in the literature and highlighted in this chapter, it can be deduced that by predisposing the expression of nutrition signaling pathways toward activating AMPK, SIRT, and IGF-1 as opposed to mTOR might bring antiaging benefits into play. This can be managed promisingly through moderate CR or by using agents able to target the aging-related signals.

References

1. Morgunova GV, Klebanov AA. Age-related AMP-activated protein kinase alterations: from cellular energetics to longevity. *Cell Biochem Funct.* 2019;37:169—176.
2. Mihaylova MM, Shaw RJ. The AMPK signalling pathway coordinates cell growth, autophagy and metabolism. *Nat Cell Biol.* 2011;13:1016—1023.
3. Lopez-Otin C, Blasco MA, Partridge L, Serrano M, Kroemer G. The hallmarks of aging. *Cell.* 2013;153:1194—1217.
4. Lovy A, Ahumada-Castro U, Bustos G, et al. Concerted action of AMPK and sirtuin-1 induces mitochondrial fragmentation upon inhibition of Ca(2+) transfer to mitochondria. *Front Cell Dev Biol.* 2020;8:378.
5. Ulgherait M, Rana A, Rera M, Graniel J, Walker DW. AMPK modulates tissue and organismal aging in a non-cell-autonomous manner. *Cell Rep.* 2014;8:1767—1780.
6. Burkewitz K, Zhang Y, Mair WB. AMPK at the nexus of energetics and aging. *Cell Metabol.* 2014;20:10—25.
7. Stenesen D, Suh JM, Seo J, et al. Adenosine nucleotide biosynthesis and AMPK regulate adult life span and mediate the longevity benefit of caloric restriction in flies. *Cell Metabol.* 2013;17:101—112.
8. Ji J, Xue TF, Guo XD, et al. Antagonizing peroxisome proliferator-activated receptor gamma facilitates M1-to-M2 shift of microglia by enhancing autophagy via the LKB1-AMPK signaling pathway. *Aging Cell.* 2018;17:e12774.
9. Lan F, Lin Y, Gao Z, Cacicedo JM, Weikel K, Ido Y. Activation of LKB1 rescues 3T3-L1 adipocytes from senescence induced by Sirt1 knock-down: a pivotal role of LKB1 in cellular aging. *Aging (Albany NY).* 2020;12:18942—18956.
10. Lee J, Tsogbadrakh B, Yang S, et al. Klotho ameliorates diabetic nephropathy via LKB1-AMPK-PGC1alpha-mediated renal mitochondrial protection. *Biochem Biophys Res Commun.* 2020;534:1040—1046.
11. Lee KY, Kim JR, Choi HC. Genistein-induced LKB1-AMPK activation inhibits senescence of VSMC through autophagy induction. *Vasc Pharmacol.* 2016;81:75—82.
12. Chen X, Zhao X, Cai H, et al. The role of sodium hydrosulfide in attenuating the aging process via PI3K/AKT and CaMKKbeta/AMPK pathways. *Redox Biol.* 2017;12:987—1003.
13. Son SM, Jung ES, Shin HJ, Byun J, Mook-Jung I. Abeta-induced formation of autophagosomes is mediated by RAGE-CaMKKbeta-AMPK signaling. *Neurobiol Aging.* 2012;33:1006 e1011—1023.
14. Mattam U, Talari NK, Paripati AK, Krishnamoorthy T, Sepuri NBV. Kisspeptin preserves mitochondrial function by inducing mitophagy and autophagy in aging rat brain hippocampus and human neuronal cell line. *Biochim Biophys Acta Mol Cell Res.* 2020;1868:118852.

15. Park SS, Seo YK, Kwon KS. Sarcopenia targeting with autophagy mechanism by exercise. *BMB Rep.* 2019;52:64–69.
16. Wang S, Kandadi MR, Ren J. Double knockout of Akt2 and AMPK predisposes cardiac aging without affecting lifespan: role of autophagy and mitophagy. *Biochim Biophys Acta, Mol Basis Dis.* 2019;1865:1865–1875.
17. Bujak AL, Crane JD, Lally JS, et al. AMPK activation of muscle autophagy prevents fasting-induced hypoglycemia and myopathy during aging. *Cell Metabol.* 2015;21:883–890.
18. Liu B, Tu Y, He W, et al. Hyperoside attenuates renal aging and injury induced by D-galactose via inhibiting AMPK-ULK1 signaling-mediated autophagy. *Aging (Albany NY).* 2018;10:4197–4212.
19. Xu TT, Li H, Dai Z, et al. Spermidine and spermine delay brain aging by inducing autophagy in SAMP8 mice. *Aging (Albany NY).* 2020;12:6401–6414.
20. Zhao H, Li T, Wang K, et al. AMPK-mediated activation of MCU stimulates mitochondrial Ca(2+) entry to promote mitotic progression. *Nat Cell Biol.* 2019;21:476–486.
21. Longo VD, Antebi A, Bartke A, et al. Interventions to slow aging in humans: are we ready? *Aging Cell.* 2015;14:497–510.
22. Cetrullo S, D'Adamo S, Tantini B, Borzi RM, Flamigni F. mTOR, AMPK, and Sirt1: key players in metabolic stress management. *Crit Rev Eukaryot Gene Expr.* 2015;25:59–75.
23. Kwon SM, Hong SM, Lee YK, Min S, Yoon G. Metabolic features and regulation in cell senescence. *BMB Rep.* 2018;52:5–12.
24. Lopez-Lluch G, Hernandez-Camacho JD, Fernandez-Ayala DJM, Navas P. Mitochondrial dysfunction in metabolism and ageing: shared mechanisms and outcomes? *Biogerontology.* 2018;19:461–480.
25. Saxton RA, Sabatini DM. mTOR signaling in growth, metabolism, and disease. *Cell.* 2017;168:960–976.
26. Zheng L, Zhang W, Zhou Y, Li F, Wei H, Peng J. Recent advances in understanding amino acid sensing mechanisms that regulate mTORC1. *Int J Mol Sci.* 2016;17.
27. Zhuang Y, Wang XX, He J, He S, Yin Y. Recent advances in understanding of amino acid signaling to mTORC1 activation. *Front Biosci.* 2019;24:971–982.
28. Kapahi P, Chen D, Rogers AN, et al. With TOR, less is more: a key role for the conserved nutrient-sensing TOR pathway in aging. *Cell Metabol.* 2010;11:453–465.
29. Kapahi P, Zid BM, Harper T, Koslover D, Sapin V, Benzer S. Regulation of lifespan in Drosophila by modulation of genes in the TOR signaling pathway. *Curr Biol.* 2004;14:885–890.
30. Wu JJ, Liu J, Chen EB, et al. Increased mammalian lifespan and a segmental and tissue-specific slowing of aging after genetic reduction of mTOR expression. *Cell Rep.* 2013;4:913–920.
31. McCloskey A, Taniguchi I, Shinmyozu K, Ohno M. hnRNP C tetramer measures RNA length to classify RNA polymerase II transcripts for export. *Science.* 2012;335:1643–1646.
32. Anisimov VN, Zabezhinski MA, Popovich IG, et al. Rapamycin extends maximal lifespan in cancer-prone mice. *Am J Pathol.* 2010;176:2092–2097.
33. Harrison DE, Strong R, Sharp ZD, et al. Rapamycin fed late in life extends lifespan in genetically heterogeneous mice. *Nature.* 2009;460:392–395.
34. Miller RA, Harrison DE, Astle CM, et al. Rapamycin-mediated lifespan increase in mice is dose and sex dependent and metabolically distinct from dietary restriction. *Aging Cell.* 2014;13:468–477.
35. Schmeisser K, Parker JA. Pleiotropic effects of mTOR and autophagy during development and aging. *Front Cell Dev Biol.* 2019;7:192.

36. Zhang N, Zhao Y. Other molecular mechanisms regulating autophagy. *Adv Exp Med Biol.* 2019;1206:261–271.
37. Nnah IC, Wang B, Saqcena C, et al. TFEB-driven endocytosis coordinates MTORC1 signaling and autophagy. *Autophagy.* 2019;15:151–164.
38. Ramsey KM, Mills KF, Satoh A, Imai S. Age-associated loss of Sirt1-mediated enhancement of glucose-stimulated insulin secretion in beta cell-specific Sirt1-overexpressing (BESTO) mice. *Aging Cell.* 2008;7:78–88.
39. Mouchiroud L, Houtkooper RH, Moullan N, et al. The NAD(+)/Sirtuin pathway modulates longevity through activation of mitochondrial UPR and FOXO signaling. *Cell.* 2013;154:430–441.
40. Yoshino J, Mills KF, Yoon MJ, Imai S. Nicotinamide mononucleotide, a key NAD(+) intermediate, treats the pathophysiology of diet- and age-induced diabetes in mice. *Cell Metabol.* 2011;14:528–536.
41. Zhang HN, Dai Y, Zhang CH, et al. Sirtuins family as a target in endothelial cell dysfunction: implications for vascular ageing. *Biogerontology.* 2020;21:495–516.
42. Sung JY, Kim SG, Kim JR, Choi HC. SIRT1 suppresses cellular senescence and inflammatory cytokine release in human dermal fibroblasts by promoting the deacetylation of NF-kappaB and activating autophagy. *Exp Gerontol.* 2021;150:111394.
43. Spannbrucker T, Ale-Agha N, Goy C, et al. Induction of a senescent like phenotype and loss of gap junctional intercellular communication by carbon nanoparticle exposure of lung epithelial cells. *Exp Gerontol.* 2018;117:106–112.
44. Wan YZ, Gao P, Zhou S, et al. SIRT1-mediated epigenetic downregulation of plasminogen activator inhibitor-1 prevents vascular endothelial replicative senescence. *Aging Cell.* 2014;13:890–899.
45. Donato AJ, Morgan RG, Walker AE, Lesniewski LA. Cellular and molecular biology of aging endothelial cells. *J Mol Cell Cardiol.* 2015;89:122–135.
46. Irie J, Itoh H. [Aging and homeostasis. Age-associated diseases and clinical application of NMN(Nicotinamide Mononucleotide).]. *Clin Calcium.* 2017;27:983–990.
47. Chen C, Zhou M, Ge Y, Wang X. SIRT1 and aging related signaling pathways. *Mech Ageing Dev.* 2020;187:111215.
48. Bradshaw PC. Acetyl-CoA metabolism and histone acetylation in the regulation of aging and lifespan. *Antioxidants.* 2021;10.
49. Mostoslavsky R, Chua KF, Lombard DB, et al. Genomic instability and aging-like phenotype in the absence of mammalian SIRT6. *Cell.* 2006;124:315–329.
50. Ota H, Akishita M, Eto M, Iijima K, Kaneki M, Ouchi Y. Sirt1 modulates premature senescence-like phenotype in human endothelial cells. *J Mol Cell Cardiol.* 2007;43:571–579.
51. Chen J, Xie JJ, Jin MY, et al. Sirt6 overexpression suppresses senescence and apoptosis of nucleus pulposus cells by inducing autophagy in a model of intervertebral disc degeneration. *Cell Death Dis.* 2018;9:56.
52. Yao H, Chung S, Hwang JW, et al. SIRT1 protects against emphysema via FOXO3-mediated reduction of premature senescence in mice. *J Clin Invest.* 2012;122:2032–2045.
53. Zu Y, Liu L, Lee MY, et al. SIRT1 promotes proliferation and prevents senescence through targeting LKB1 in primary porcine aortic endothelial cells. *Circ Res.* 2010;106:1384–1393.
54. Kanfi Y, Naiman S, Amir G, et al. The sirtuin SIRT6 regulates lifespan in male mice. *Nature.* 2012;483:218–221.
55. Kawahara TL, Michishita E, Adler AS, et al. SIRT6 links histone H3 lysine 9 deacetylation to NF-kappaB-dependent gene expression and organismal life span. *Cell.* 2009;136:62–74.

56. Wong DW, Soga T, Parhar IS. Aging and chronic administration of serotonin-selective reuptake inhibitor citalopram upregulate Sirt4 gene expression in the pre-optic area of male mice. *Front Genet.* 2015;6:281.
57. Gomes AP, Price NL, Ling AJ, et al. Declining NAD(+) induces a pseudohypoxic state disrupting nuclear-mitochondrial communication during aging. *Cell.* 2013;155:1624−1638.
58. Chang HC, Guarente L. SIRT1 mediates central circadian control in the SCN by a mechanism that decays with aging. *Cell.* 2013;153:1448−1460.
59. Kida Y, Goligorsky MS. Sirtuins, cell senescence, and vascular aging. *Can J Cardiol.* 2016;32:634−641.
60. Potente M, Ghaeni L, Baldessari D, et al. SIRT1 controls endothelial angiogenic functions during vascular growth. *Genes Dev.* 2007;21:2644−2658.
61. van der Veer E, Ho C, O'Neil C, et al. Extension of human cell lifespan by nico-tinamide phosphoribosyltransferase. *J Biol Chem.* 2007;282:10841−10845.
62. Hong W, Mo F, Zhang Z, Huang M, Wei X. Nicotinamide mononucleotide: a promising molecule for therapy of diverse diseases by targeting NAD+ metabolism. *Front Cell Dev Biol.* 2020;8:246.
63. Klimova N, Long A, Kristian T. Nicotinamide mononucleotide alters mitochondrial dynamics by SIRT3-dependent mechanism in male mice. *J Neurosci Res.* 2019;97:975−990.
64. Ratajczak J, Joffraud M, Trammell SA, et al. NRK1 controls nicotinamide mono-nucleotide and nicotinamide riboside metabolism in mammalian cells. *Nat Commun.* 2016;7:13103.
65. Fletcher RS, Ratajczak J, Doig CL, et al. Nicotinamide riboside kinases display redundancy in mediating nicotinamide mononucleotide and nicotinamide riboside metabolism in skeletal muscle cells. *Mol Metabol.* 2017;6:819−832.
66. Yamaguchi S, Yoshino J. The pathophysiological importance and therapeutic potential of NAD' biosynthesis and mitochondrial sirtuin SIRT3 in age-associated diseases. *Nihon Rinsho.* 2016;74:1447−1455.
67. Yoshino J, Baur JA, Imai SI. NAD(+) intermediates: the biology and therapeutic potential of NMN and NR. *Cell Metabol.* 2017;27:513−528.
68. Braidy N, Berg J, Clement J, et al. Role of nicotinamide adenine dinucleotide and related precursors as therapeutic targets for age-related degenerative diseases: rationale, biochemistry, pharmacokinetics, and outcomes. *Antioxidants Redox Signal.* 2018;30:251−294.
69. Mendelsohn AR, Larrick JW. The NAD+/PARP1/SIRT1 Axis in aging. *Rejuvena-tion Res.* 2017;20:244−247.
70. Zhang N, Sauve AA. Regulatory effects of NAD(+) metabolic pathways on sirtuin activity. *Prog Mol Biol Transl Sci.* 2018;154:71−104.
71. Wennberg AMV, Hagen CE, Petersen RC, Mielke MM. Trajectories of plasma IGF-1, IGFBP-3, and their ratio in the mayo clinic study of aging. *Exp Gerontol.* 2018;106:67−73.
72. Arwert LI, Veltman DJ, Deijen JB, van Dam PS, Drent ML. Effects of growth hor-mone substitution therapy on cognitive functioning in growth hormone deficient patients: a functional MRI study. *Neuroendocrinology.* 2006;83:12−19.
73. Deijen JB, Arwert LI, Drent ML. The GH/IGF-I Axis and cognitive changes across a 4-year period in healthy adults. *ISRN Endocrinol.* 2011:249421, 2011.
74. Bozdagi O, Tavassoli T, Buxbaum JD. Insulin-like growth factor-1 rescues synaptic and motor deficits in a mouse model of autism and developmental delay. *Mol Autism.* 2013;4:9.
75. Molina DP, Ariwodola OJ, Weiner JL, Brunso-Bechtold JK, Adams MM. Growth hormone and insulin-like growth factor-I alter hippocampal excitatory synaptic transmission in young and old rats. *Age (Dordr).* 2013;35:1575−1587.

76. Tarantini S, Nyul-Toth A, Yabluchanskiy A, et al. Endothelial deficiency of insulin-like growth factor-1 receptor (IGF1R) impairs neurovascular coupling responses in mice, mimicking aspects of the brain aging phenotype. *Geroscience*. 2021;43:2387–2394.
77. Tarantini S, Balasubramanian P, Yabluchanskiy A, et al. IGF1R signaling regulates astrocyte-mediated neurovascular coupling in mice: implications for brain aging. *Geroscience*. 2021;43:901–911.
78. Tarantini S, Tran CHT, Gordon GR, Ungvari Z, Csiszar A. Impaired neurovascular coupling in aging and Alzheimer's disease: contribution of astrocyte dysfunction and endothelial impairment to cognitive decline. *Exp Gerontol*. 2016;94:52–58.
79. Toth P, Tarantini S, Ashpole NM, et al. IGF-1 deficiency impairs neurovascular coupling in mice: implications for cerebromicrovascular aging. *Aging Cell*. 2015;14:1034–1044.
80. Junnila RK, List EO, Berryman DE, Murrey JW, Kopchick JJ. The GH/IGF-1 axis in ageing and longevity. *Nat Rev Endocrinol*. 2013;9:366–376.
81. Ewald CY, Castillo-Quan JI, Blackwell TK. Untangling longevity, dauer, and healthspan in *Caenorhabditis elegans* insulin/IGF-1-signalling. *Gerontology*. 2017;64:96–104.
82. Colman RJ, Anderson RM, Johnson SC, et al. Caloric restriction delays disease onset and mortality in rhesus monkeys. *Science*. 2009;325:201–204.
83. Colman RJ, Beasley TM, Kemnitz JW, Johnson SC, Weindruch R, Anderson RM. Caloric restriction reduces age-related and all-cause mortality in rhesus monkeys. *Nat Commun*. 2014;5:3557.
84. Civitarese AE, Carling S, Heilbronn LK, et al. Calorie restriction increases muscle mitochondrial biogenesis in healthy humans. *PLoS Med*. 2007;4:e76.
85. Larson-Meyer DE, Heilbronn LK, Redman LM, et al. Effect of calorie restriction with or without exercise on insulin sensitivity, beta-cell function, fat cell size, and ectopic lipid in overweight subjects. *Diabetes Care*. 2006;29:1337–1344.
86. Johnson ML, Distelmaier K, Lanza IR, et al. Mechanism by which caloric restriction improves insulin sensitivity in sedentary obese adults. *Diabetes*. 2016;65:74–84.
87. Heilbronn LK, de Jonge L, Frisard MI, et al. Effect of 6-month calorie restriction on biomarkers of longevity, metabolic adaptation, and oxidative stress in overweight individuals: a randomized controlled trial. *JAMA*. 2006;295:1539–1548.
88. Ahmed T, Das SK, Golden JK, Saltzman E, Roberts SB, Meydani SN. Calorie restriction enhances T-cell-mediated immune response in adult overweight men and women. *J Gerontol A Biol Sci Med Sci*. 2009;64:1107–1113.
89. Fontana L, Partridge L. Promoting health and longevity through diet: from model organisms to humans. *Cell*. 2015;161:106–118.
90. Riera CE, Merkwirth C, De Magalhaes Filho CD, Dillin A. Signaling networks determining life span. *Annu Rev Biochem*. 2016;85:35–64.
91. Madeo F, Carmona-Gutierrez D, Hofer SJ, Kroemer G. Caloric restriction mimetics against age-associated disease: targets, mechanisms, and therapeutic potential. *Cell Metabol*. 2019;29:592–610.
92. Petrovski G, Das DK. Does autophagy take a front seat in lifespan extension? *J Cell Mol Med*. 2010;14:2543–2551.
93. Rachakatla A, Kalashikam RR. Calorie restriction-regulated molecular pathways and its impact on various age groups: an overview. *DNA Cell Biol*. 2022;41:459–468.
94. Rothschild J, Hoddy KK, Jambazian P, Varady KA. Time-restricted feeding and risk of metabolic disease: a review of human and animal studies. *Nutr Rev*. 2014;72:308–318.
95. Anton SD, Moehl K, Donahoo WT, et al. Flipping the metabolic switch: understanding and applying the health benefits of fasting. *Obesity*. 2018;26:254–268.

96. de Cabo R, Mattson MP. Effects of intermittent fasting on health, aging, and disease. *N Engl J Med.* 2019;381:2541−2551.
97. Patterson RE, Sears DD. Metabolic effects of intermittent fasting. *Annu Rev Nutr.* 2017;37:371−393.
98. Villareal DT, Fontana L, Weiss EP, et al. Bone mineral density response to caloric restriction-induced weight loss or exercise-induced weight loss: a randomized controlled trial. *Arch Intern Med.* 2006;166:2502−2510.
99. Fontana L, Weiss EP, Villareal DT, Klein S, Holloszy JO. Long-term effects of calorie or protein restriction on serum IGF-1 and IGFBP-3 concentration in humans. *Aging Cell.* 2008;7:681−687.
100. Fontana L, Klein S, Holloszy JO. Effects of long-term calorie restriction and endurance exercise on glucose tolerance, insulin action, and adipokine production. *Age (Dordr).* 2010;32:97−108.
101. Wang L, Karpac J, Jasper H. Promoting longevity by maintaining metabolic and proliferative homeostasis. *J Exp Biol.* 2014;217:109−118.

CHAPTER 4

Dysregulated proteostasis: mechanisms and links to aging

Yasmeen Al-Mufti[a], Stephen Cranwell[a] and Rahul S. Samant

Signalling Programme, The Babraham Institute, Cambridge, United Kingdom

1. Introduction

A PubMed search for "proteostasis," a portmanteau of "protein homeostasis," flags its first appearance in the biomedical literature in 2008.[1] It is therefore remarkable that loss of proteostasis was recognized as a primary hallmark of aging in the landmark report by Lopez-Otin et al. only 5 years later.[2] Important to this recognition were a series of influential essays and review articles in the intervening years that synthesized decades of progress from diverse disciplines into a compelling conceptual framework centered on the presence of a coordinated cellular network dedicated to maintaining proteome balance and robustness throughout an organism's life span, which goes awry during aging[3–9]. Despite several subsequent reevaluations and extensions of the original aging hallmarks,[10–14] loss of proteostasis remains acknowledged as a fundamental process at the heart of aging-associated functional decline.

According to the Human Protein Atlas (https://www.proteinatlas.org/, v21.1), the human protein-coding genome contains 8000—9000 housekeeping genes and another 10,000—11,000 tissue-specific genes.[15] Although estimates based on experimental observations vary, it is generally accepted that most human cells contain well over a billion (10^9) protein molecules at any given time,[16,17] with an apparent dynamic range (i.e., the difference in concentration between the least and most abundant protein) of seven orders of magnitude.[18] As it emerges from a translating ribosome as a linear chain of amino acids, each molecule must: (1) fold into its native three-dimensional structure, (2) travel to the correct subcellular location(s), and (3) find its physiologic binding partners—all while avoiding erroneous interactions with billions of other macromolecules in the surrounding milieu. Consistent failures in any of these steps can have disastrous

[a] These authors contributed equally.

Molecular, Cellular, and Metabolic Fundamentals of Human Aging
ISBN 978-0-323-91617-2
https://doi.org/10.1016/B978-0-323-91617-2.00004-3

© 2023 Elsevier Inc.
All rights reserved.

consequences for the organism as a whole, not only because of the loss of that protein's required function at the time, but also owing to gain-of-function toxicity associated with misfolded or mislocalized proteins interfering with critical cellular processes. Similar interference can be caused by dysregulation of a protein's expression or degradation rates, which perturb steady-state proteome dynamics. Furthermore, widespread remodeling of the proteome is required during several developmental, differentiation, and stress response programs. Because almost all diseases for which age is the major risk factor are linked mechanistically to aberrant accumulation of misfolded or mislocalized proteins,[19,20] organisms have evolved specific machineries to regulate every stage of a protein's life span—from initial synthesis, folding, targeting, and assembly to conformational maintenance in changing physiologic conditions, to timely clearance (e.g., when the protein has fulfilled its function or is irreparably damaged). The collective and coordinated action of these machineries is central to healthy proteostasis.

2. Functional modules of the proteostasis network

The remit of what qualifies as a proteostasis-related process—and, by extension, the composition of the proteostasis network (PN)—is necessarily broad. At the last count, approximately 2000 proteins were annotated in the core PN of humans, separated into three modules: synthesis, folding, and degradation (Fig. 4.1).[21] Yet this classification schema excludes several auxiliary protein classes with a vital role in cellular protein homeostasis (e.g., factors regulating intracellular targeting, protein import and export, and protein complex assembly). It also ignores proteins involved in transcriptional modulation of the proteostatic balance, such as stress response factors. All of these components are included in the original definition of proteostasis: "controlling the concentration, conformation, binding interactions (quaternary structure), and location of individual proteins making up the proteome by readapting the innate biology of the cell, often through transcriptional and translational changes."[1]

2.1 Synthesis module

Most protein synthesis takes place through cytoplasmic ribosomes, one or more of which assemble on a messenger RNA (mRNA) transcript to initiate translation at roughly 5—6 amino acids/s.[22] Because biological processes are inherently error-prone, quality control (QC) mechanisms exist

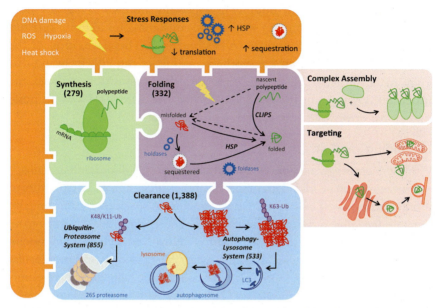

Figure 4.1 *Functional module of the proteostasis network (PN). The core PN consists of modules for protein synthesis, folding (either de novo via the chaperones linked to protein synthesis [CLIPS] or stress-induced via heat shock proteins [HSP]), and clearance (via the ubiquitin–proteasome system or autophagy–lysosome system). In eukaryotes, the PN is extended with modules to help proteins navigate through a crowded three-dimensional environment to locate their binding partners or arrive at the correct subcellular compartment(s). In both prokaryotes and eukaryotes, PN composition can be remodeled rapidly and reversibly in response to various proteotoxic stresses. The number of proteins annotated in each core PN module in humans is indicated in parentheses, according to Klaips et al.[21]* mRNA, messenger RNA; ROS, reactive oxygen species.

at almost every step to maintain fidelity. Protein translation is no different. Therefore, the more than 250 proteins in the synthesis module of the human PN are composed of not only ribosomal proteins and translation factors, but also an assortment of QC factors that ensure translational fidelity—from the mRNA being read to the protein emerging from the ribosome exit tunnel.[23–26]

The amino acid misincorporation rate for protein translation is around 1 in 10,000.[27,28] Given the distribution of protein lengths and abundances in human cells, an estimated 5%–15% of proteins will contain an erroneous amino acid in their sequence. Consequently, a considerable proportion of the nascent proteome is cotranslationally degraded (estimates vary between

12% and 30%).[29–32] This number is especially noteworthy given how energetically expensive protein synthesis is, consuming up to three-quarters of a cell's adenosine triphosphate (ATP) budget.[33] Therefore, most of the cell's energy is spent within the synthesis module of the PN. Ensuring that newly made proteins are of the highest quality presumably reduces the burden on downstream PN modules, leaving them free to focus on post-translational errors that arise during a protein's life span.

2.2 Folding module

With the increasingly notable exception of intrinsically disordered proteins,[34] most proteins must be folded into a specific three-dimensional (3D) structure to be fully functional. This correct native structure is specified by the primary sequence (i.e., the order of amino acids in the polypeptide chain). Small, globular proteins can fold unassisted in vitro.[35] However, with the evolution of larger, multidomain proteins, together with the crowded macromolecular environment of a cell (20–30% of a cell's volume is occupied by macromol-ecules, which equates to an effective concentration of 200–300 mg/mL[36]), folding in vivo necessitates the presence of a set of accessory proteins known as molecular chaperones.[37–39] The chaperome is composed of a diverse set of proteins unified by their ability to favor the native folded state and/or disfavor nonnative misfolded states of their substrates (commonly referred to as clients). At the most fundamental level, a molecular chaperone binds surface-exposed hydrophobic sequence elements within a client protein—elements that would normally be buried, such as within its core (for globular domains) or the lipid bilayer (for transmembrane segments)—and protects the polypeptide chain from nonproductive, off-pathway interactions with itself or other chains. In addition to this ability to protect their clients from aggregation, referred to as a chaperone's holdase function, many chaperones act as foldases, coupling ATP binding and hydrolysis cycles to accelerate folding reactions, and/or to unfold nonnative, kinetically trapped states.[40,41] ATP-coupled unfolding is also important for the disentanglement of intermolecular misfolded protein species (such as oligomers or aggregates)—although this disaggregase ability is restricted to a smaller subset of chaperones. Taken together, these holdase, foldase, unfoldase, and disaggregase activities provide the folding module with the tool kit required to help any unfolded or misfolded protein species it encounters navigate through the thermodynamic landscape (Fig. 4.2).

In addition to distributing the various chaperoning activities among different members of the folding module, eukaryotic cells have divided labor through the evolution of two partially overlapping chaperone

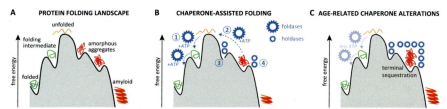

Figure 4.2 *The chaperone-assisted folding landscape in youth and aging. (A) An unfolded protein needs to reach its native folded state through a complex energetic landscape involving on-pathway yet kinetically trapped folding intermediates, as well as off-pathway intramolecular and intermolecular energetic minima. Amyloid fibrils are especially problematic because they occupy a lower free energy state than the native folded conformation. (B) Molecular chaperones bind unfolded proteins and help them navigate the landscape by (1) promoting on-pathway folding steps; (2) unfolding and/or disaggregating off-pathway conformations; (3) simplifying energetic topology by limiting access to local kinetic traps; and (4) preventing aggregates from entering amyloid and other fibrillar conformations. Functions 1 and 2 are ATP-dependent, whereas 3 and 4 are ATP-independent. (C) During aging, a decline in available ATP necessitates a shift in resource allocation to focus on folding of newly synthesized unfolded polypeptides rather than trying to disaggregate off-pathway species; these are instead terminally sequestered by holdases (e.g., small heat shock proteins).*

networks: chaperones linked to protein synthesis (CLIPS) for de novo folding, and heat shock protein (HSP) chaperones for conformational maintenance and misfolded protein triage.[42] HSP components are transcriptionally upregulated in response to protein misfolding (proteotoxic) stresses; by contrast, CLIPS chaperones as a whole are transcriptionally repressed, mirroring the global repression of protein translation during stress. Separating the chaperome in this way allows eukaryotic cells to boost PN capacity by selectively increasing levels of only chaperones able to bind and triage misfolded states, while repressing all nonessential protein synthesis that could otherwise increase the proteostasis burden.

2.2.1 De novo folding

As opposed to folding a chemically denatured protein in vitro, a newly synthesized protein in vivo must be folded as it emerges from the ribosome vectorially, N-terminal first.[43] The ribosome exit tunnel—from the peptidyl transferase center where amino acids are added to the growing chain to the exit port where the chain emerges from the ribosome—is approximately 100 Å long and 20 Å wide. This tunnel is therefore too narrow to accommodate most structural elements (outside alpha helices and simple tertiary folds), although some small protein domains (<50 amino acids) appear to be able to fold fully within the exit tunnel.[44–46] Far from being a passive channel through which the polypeptide must travel on its way to the cytoplasmic chaperone machinery, interactions between the exit tunnel and the nascent chain have a key role in regulating folding.[47,48]

If not fully folded within the exit tunnel, the nascent polypeptide emerging from the ribosome is engaged sequentially by the CLIPS network, which consists of an evolutionarily conserved series of molecular chaperones and co-chaperones that direct the translating chain through the energetic folding landscape to its native 3D structure.[38,49,50] The first of these bind the polypeptide on the ribosome itself: ribosome-associated complex and nascent chain-associated complex. At least 30% of the proteome requires further assistance beyond the ribosome; these polypeptides are transferred to HSP70, which functions together with an assortment of co-chaperones to facilitate folding of its clients through ATP-coupled binding-and-release cycles.[51–53] HSP70 also acts as a central hub for the transfer of some clients to more specialist chaperones, such as the ring-shaped chaperonin T-complex protein ring complex (TRiC, also known as CCT, chaperonin containing TCP-1) ($\sim$ 10% of the proteome)[54] or the HSP90 system ($\sim$ 300 client proteins).[55] Therefore, different proteins will exit the CLIPS/de novo folding module at different points, depending on their specific structural intricacies. Regardless of the exact point at which a protein reaches its native structure, the chaperome has evolved to ensure that nascent chains emerging from the ribosome are protected from exposure to nonproductive interactions with the cytoplasmic milieu until they are less vulnerable to misfolding and aggregation.

2.2.2 Conformational maintenance and misfolded protein triage

Even once proteins reach their native state, they are not safe from misfolding. It had been assumed that the native folded structure of a protein represented its most thermodynamically stable conformation under physiologic conditions, and that the observed formation of even more stable amyloid fibrils occurred only at much higher concentrations (e.g., under pathophysiologic conditions in which unfolded or misfolded protein conformations accumulate well above their steady-state levels). However, it is becoming increasingly clear that amyloid fibrils may be the most thermodynamically stable state for a considerable proportion of the proteome even under physiologic conditions.[56] Combined with the fact that other off-pathway misfolded states and on-pathway folding intermediates are only marginally less stable than the correctly folded structure, it is likely that the native proteome is inherently metastable in perfectly healthy cells and is maintained in a functional, nonamyloid state by kinetic barriers rather than thermodynamic ones.[57,58]

Because of the dangers posed by stress-induced misfolding—especially for the nascent and metastable proteomes—all cells have evolved a variety of stress responses that rapidly and robustly boost their proteostatic capacity. In fact, HSPs were initially discovered and named as a result of their greatly increased expression during heat shock,[59] a stress that has a considerable impact on the folding landscape and leads to increased off-pathway misfolded and aggregated states. The HSP70 and HSP90 systems acts as central hubs for protein QC, working individually or in collaboration to recognize exposed hydrophobic sequence stretches within misfolded proteins and triaging them for refolding or clearance. A distinct set of co-chaperones exists for HSP70 versus HSP90, with the co-chaperone Sti1/HOP mediating transfer between the two systems. These co-chaperones ($\sim$50 for each chaperone in humans) differ in their client and tissue specificity, and function as adaptors that provide versatility to the systems[60,61]—an important characteristic in PQC, because any protein in the proteome has the potential to misfold. This, combined with the fact that variants of this machinery exist in most subcellular compartments, ensure that misfolded proteins can be triaged rapidly in situ regardless of when and where they arise.

In contrast to the diverse functions of the HSP70 and HSP90 machineries, small HSPs (sHSPs) have no known function in de novo folding and no foldase activity. Although members of the sHSP family vary greatly in structure, expression patterns (both constitutively and under stress), tissue specificities, and client preferences, they are all ATP-independent holdases that cycle between oligomeric states to sequester their clients away spatially from the cellular milieu.[62,63] Importantly, sHSP-mediated sequestration of misfolded proteins is proposed to serve a cytoprotective function by preventing the accumulation of alternative misfolded conformations that have potentially higher gain-of-function toxicity.

2.2.3 Spatial sequestration and resolution

Cytoprotective spatial sequestration, such as that mediated by sHSPs, is a rapid response to a variety of proteotoxic stresses.[64,65] However, it does not explicitly require acute stress response induction to be triggered; rather, it exists as a surveillance system even under physiologic conditions. The accumulation of misfolded protein species above a certain threshold, indicating that the PQC machinery may be overwhelmed, leads to their sequestration into Q-bodies (also known as CytoQ or stress foci).[66] Q-bodies are sites of active PQC and triage to refolding or clearance pathways. They are rich in both ATP-dependent (HSP70—HSP90) and independent (sHSPs and other sequestration factors) chaperones, as well as components of the ubiquitin—proteasome clearance system (see the next section). Therefore, transient and/or local increases in the misfolded protein load can be resolved in situ through this evolutionarily conserved sequestration-and-triage mechanism.

If these Q-bodies cannot be resolved in situ, they are actively transported and eventually coalesce into a single juxtanuclear inclusion, varyingly referred to as the juxtanuclear quality control compartment (JUNQ), intranuclear QC compartment (INQ), or aggresome, depending on the exact composition, location, and cellular context.[67] Regardless of the nomenclature, this juxtanuclear site serves as a centralized depot for misfolded proteins in situations in which the active, ATP-dependent PQC machinery cannot keep up with the genesis of substrates that need triage. Centralizing PQC allows the PN to simplify triage of misfolded proteins detected by multiple surveillance systems to a single site for downstream processing. Therefore, in addition to being rich in both holdases and foldases, cargoes from juxtanuclear inclusions have been shown to be targeted to almost all of the clearance mechanisms outlined in the next section, including the ubiquitin—proteasome system, autophagy—lysosome system, terminal sequestration, and asymmetric inheritance.

A key feature of spatial sequestration in response to acute stresses is its reversibility (i.e., the potential for resolution of the sequestration site by either refolding or clearance once the stress has passed and homeostasis is restored). Therefore, these sites—Q-bodies, JUNQ/INQ, and the aggresome—can be seen as temporary quarantine zones for situations in which the cellular PQC capacity has been exceeded. Here, misfolded proteins are kept separate from the folded and functional protein population until PQC machineries are available again for active triage. Without this mechanism, even low levels of misfolded proteins floating in the cellular milieu could seed further misfolding and aggregation, especially of nascent and metastable proteomes.

2.3 Clearance module

Of the almost 2000 proteins in the human core PN, over two-thirds are components of the degradation module (Fig. 4.1).[21] The large proportion dedicated to this process highlights the importance of ensuring timely turnover of proteins after they have fulfilled their specific function (e.g., extinguishing a signaling output, or progressing through cell cycle phases and developmental programs). The number could also be higher than for the other network modules, because (1) proteins annotated within the degradation module include hundreds of specific adaptors for individual substrates; (2) many ubiquitylation enzymes in the module likely have no or limited roles in degradation, but function to form nonproteolytic ubiquitin chains; and (3) the degradation module (especially the autophagy—lysosome system) includes factors that process molecules other than proteins, such as lipids, carbohydrates, organelles, and metabolites.

In addition to this programmed degradation—which occurs on functional, native proteins—the degradation module needs to be able to detect nonfunctional, defective, or damaged proteins, either during their biogenesis or at any point in their life span. This also applies to misfolded proteins that cannot be refolded. Although factors governing the refolding versus degradation triage decision are various and still not fully understood, they are likely related to the stochastic binding-and-release cycles that drive client protein folding and maturation. Too many nonproductive folding cycles for the same client molecule would favor its targeting to the degradative machinery.[61] Competition between co-chaperones with different on- and off-rates for the central chaperone and/or client help mediate triage. Examples include de novo folding and conformational maintenance of cytoplasmic clients of the HSP70-HSP90 systems[68] and membrane protein insertion in the correct organelle.[69,70] Competition between cofactors of differing affinities also has a role in targeting to different systems within the degradation module (e.g., proteasome versus autophagy).[71,72] Linking triage to chaperone residence time thus prevents fruitless titration of the PQC machinery by providing an off-ramp for refractory client species.

2.3.1 Ubiquitin—proteasome system

The preferred machinery for degradation of most of the cellular proteome is the proteasome.[73,74] The catalytic unit of the proteasome is the 20S core particle, which has a stacked barrel-shaped structure with two hetero-heptameric beta rings sandwiched between two hetero-heptameric alpha rings (i.e., an $\alpha_{1-7}{:}\beta_{1-7}{:}\beta_{1-7}{:}\alpha_{1-7}$ conformation). Access to the catalytic subunits in the inner beta rings of the 20S core particle is gated by flexible tails at the N-terminal of each alpha subunit. This gate can be opened by docking of a regulatory particle to form the 26S proteasome: a 2.5-MDa machine consisting of a 20S core bound at one or both ends by 19S regulatory particles.

The 19S regulatory particle consists of a base that can open the 20S gate in an ATP-dependent manner and a lid that contains subunits to manage substrate recognition and preprocessing to make them competent for proteolytic cleavage through the 20S core. Proteins can thus be targeted to the 26S proteasome by the posttranslational attachment of ubiquitin—a 76—amino acid (8-kDa) globular protein that is highly conserved in all eukaryotes. Although it was initially proposed that ubiquitin tagging of protein molecules invariably signaled their proteasomal targeting and degradation, it has since become apparent that ubiquitylation is a posttranslational modification with a diverse range of nonproteasomal—and nonproteolytic—signaling roles.[75]

The diversity of ubiquitin's signaling outputs stems from its ability to polymerize through any of its seven internal lysine residues or its N-terminal methionine. Polyubiquitin chains formed from different lysine linkages specify different downstream fates for the protein to which they are attached under different contexts, generating a ubiquitin code.[76,77] The most common signal for proteasomal degradation is a chain of four or more ubiquitin moieties linked through the lysine-48 residue in each ubiquitin (K48-ubiquitin), although chains with other linkages are involved in specific contexts. For example, branched ubiquitin chains linked by both K48 and K11 appear to be preferential signals for proteasomal targeting of misfolded proteins.[78—80]

Ubiquitin attachment, both initially onto the substrate and onto existing ubiquitylated substrates to extend a chain, is an intricate process regulated by a suite of ubiquitin activating enzymes (E1), ubiquitin conjugating enzymes (E2), and ubiquitin ligases (E3).[81] Ubiquitin chains can also be removed (one moiety at a time, or all at once) by de-ubiquitylases (DUBs). Humans possess two E1s, 30—50 E2s, and over 600 E3s, as well as over 100 DUBs. Therefore, E3s and DUBs, the frontline enzymes that confer substrate specificity to the ubiquitin—proteasome system, represent at least 5% of human protein-coding genes. Add to this number the readers of ubiquitin-modified substrates (ubiquitin-binding proteins and receptors), as well as several ubiquitin-like proteins (e.g., SUMO, NEDD8, ISG15, ATG8/LC3) with their own enzymes and readers that are still being cataloged and characterized,[82] and it quickly becomes clear that the cell dedicates considerable resources to regulating the ubiquitin code.

With regard to misfolded protein triage and PQC, the handover from synthesis or folding modules to the degradation module occurs most commonly through recruitment of an E3 ubiquitin ligase. Importantly, routes must exist to target proteins for degradation at every major checkpoint in their life cycle. One of the fastest-developing areas in the proteostasis field is that of ribosome QC (RQC), through which a nascent chain is detected on stalled ribosomes and K48-ubiquitylated by listerin/ Ltn1 for proteasomal targeting and degradation.[23–25] Away from the ribosome, molecular chaperones—especially HSP70 and HSP90—provide platforms for E3 ubiquitin ligase recruitment. The best characterized of these is C-terminal of HSC70-interacting protein (CHIP), an E3 ligase that can bind the C-terminal tail of either HSP70 or HSP90 via a tetra-tricopeptide repeat (TPR) motif.[83] Because TPR motifs are common among co-chaperones of the HSP70-HSP90 machinery, the fate of client—Chapter 1 complexes can be tuned at multiple levels by co-chaperone competition at the C-terminal tail, according to the appropriate conformational (e.g., based on how difficult each individual nonnative species is to fold) and cellular (e.g., based on relative stress on the folding versus degradation machineries) cues.

Ubiquitin-mediated proteasomal degradation is also employed for membrane proteins (e.g., endoplasmic reticulum [ER]–associated degradation [ERAD], in which an intrinsic membrane-spanning E3 ubiquitin ligase detects misfolded domains in the lumen, transmembrane, or cytosolic face of an ER membrane protein [ERAD-L, -M, or −C]).[84] Ubiquitylated membrane proteins are then extracted from the membrane by the ubiquitin-binding, ring-shaped complex p97/VCP, after which they can be degraded by the proteasome.[85] The ERAD-L pathway is also employed for misfolded nonmembrane proteins in the ER lumen (e.g., secreted proteins such as hormones). Such globular misfolded proteins must also be retro-translocated through the ER membrane by p97/VCP, because neither ubiquitin nor proteasomes exist in the ER lumen. p97/VCP has diverse roles in proteostasis as an ATP-dependent extractase of ubiquitylated proteins, such as from chromatin, stalled ribosomes (for RQC), protein aggregates, and other membranes in addition to the ER.[86]

2.3.2 Autophagy—lysosome system

Whereas the proteasome is a machine for degrading individually unfolded protein chains, the lysosome is an acidic subcellular compartment that degrades much more indiscriminately through the action of various hydrolases for proteins and other macromolecules (e.g., nucleic acids, carbohydrates, lipids).[87] Therefore, lysosomes are major recycling centers for most cellular matter and have a central role in signaling and metabolism.[88]

Pathways that send different cellular matter to the lysosome are necessarily diverse. The best characterized of these is autophagy (Greek: "self-eating"), a process in which a membrane segment conjugated with the ubiquitin-like protein ATG8/LC3 surrounds the material to be degraded before fusing with the lysosomal membrane, releasing the material to the hydrolases in it. Autophagy can act nonselectively, such as under bulk macroautophagy under nutrient-limiting conditions, in which a double-membraned structure (derived from the ER and other organelles) called the autophagosome engulfs and targets to the lysosome entire cytoplasmic regions that consequently serve as a source for replenishing macromolecular building blocks and nutrient stores. In selective autophagy, on the other hand, specific macromolecular cargoes (e.g., misfolded protein aggrephagy), organelles (e.g., mitophagy, ER-phagy), or pathogens (xenophagy) are targeted to autophagic membrane structures via interactions between ubiquitin moieties on the cargo and ATG8/LC3 on the membranes. Cargo—LC3 interactions are bridged by a handful of selective autophagy cargo receptors that contain both a ubiquitin-binding domain (UBA, UBZ, or UBAN) and an LC3-interacting region.[89] In general, all selective autophagy receptors prefer ubiquitin chains with a more open conformation (e.g., K63-Ub and linear M1-Ub) as opposed to the more closed chains that signal proteasomal targeting (e.g., K48-Ub, K11-Ub).[90]

The six selective autophagy cargo receptors in humans have both overlapping and distinct functions, contexts, and substrates—highlighted by a systematic proteomics analysis.[91] For example, although p62/SQSTM1 is the most extensively studied selective autophagy cargo receptor for misfolded proteins, the related receptors NDP52, TOLLIP, optineurin, and TDP52 have also been implicated in protein aggregate clearance to varying extents.[92—95]

Besides macroautophagy—nonselective or selective—several other routes exist to target cargoes to the lysosome. Some of these (e.g., microautophagy, chaperone-mediated autophagy) act at a single-molecule level (i.e., more like proteasomal degradation). Others exist for endocytosed membrane receptors via late endosomes/multivesicular bodies. Extracellular cargoes can also be endocytosed via pathways such as LC3-associated phagocytosis,[96] LC3-associated endocytosis,[97] and chaperone-and receptor-mediated extracellular protein degradation.[98]

2.3.3 Terminal sequestration

The preferred route for turnover of proteins involves the ubiquitin—proteasome and autophagy—lysosome systems. However, alternative pathways exist that lead to terminal sequestration of certain misfolded proteins in a peripheral cytoplasmic inclusion, sometimes referred to as the insoluble protein deposit (IPOD).[99] As the name implies, the IPOD is biophysically and functionally distinct from dynamic and reversible inclusions such as Q-bodies or the JUNQ/aggresome. It is a site at the cellular periphery for depositing prions and other amyloidogenic proteins that are otherwise resistant to degradation by the usual clearance machineries. In fact, a major proposed mechanism for toxicity of misfolded neurodegeneration-associated proteins is the titration of PQC machineries such that the PN capacity is lowered and rendered extremely vulnerable to additional stresses.[100,101] An example of this is the clogging up of 20S proteasomes by neurodegeneration-associated misfolded proteins (e.g., amyloid-beta oligomers in Alzheimer's disease), which leads to the accumulation of proteasomal substrates that would be routinely degraded under physiologic conditions.[102] Although not a degradation pathway as such, terminal sequestration at the peripheral IPOD serves a similar clearance function in that it removes aggregation-prone proteins from the cellular milieu but avoids the potential for additional stress on the protein degradation machineries, leaving them free to serve their roles in physiologic protein turnover.

2.3.4 Other clearance mechanisms

Dividing cells have additional possibilities for clearing their misfolded and damaged proteome. Asymmetric cell divisions—most commonly studied in the unicellular budding yeast *Saccharomyces cerevisiae*, but relevant in most organisms, including in maintaining somatic stem cell niches—are generally accompanied by asymmetric inheritance of damage.[103–105] For example, in budding yeast, the mother cell retains damaged organelles and protein aggregates (including the JUNQ and IPOD), ensuring that the daughter cell is as damage-free as possible. Such asymmetric inheritance is also likely to exist for proteostasis machineries themselves, as demonstrated for lysosomes and autophagic compartments in hematopoietic stem cells.[106] The JUNQ and IPOD are also asymmetrically inherited in otherwise symmetrically dividing mammalian cell lines in culture, although not as efficiently as in yeast.[107] Retaining damage in one cell over another is likely a crucial part of the rejuvenating properties of cell division, promoting fitness at a population level—especially under conditions of stress.[105,108,109]

An alternative strategy for rapidly dividing cells is to dilute damage rather than segregating it, as is the case for dermal fibroblasts that symmetrically divide as part of the wound-healing response.[110] In such fibroblasts, long-lived proteins (i.e., those whose half-life is longer than the division time of the cell) need not be actively degraded as long as the fibroblasts are dividing fast enough for no single cell to accumulate too large of a damaged protein burden. Once the wound-healing response is terminated, the fibroblasts revert to a dormant, quiescent state. In this situation, the proteostasis strategy for the same long-lived proteins changes to lysosomal targeting via autophagy. This is an example of context-dependent proteostasis, with different strategies employed for the same substrates in the same cell population, depending on the proliferative status of the cell.

Although terminally differentiated cells cannot employ division-based asymmetric segregation or dilution to meet their PQC needs, they may nevertheless reduce intracellular misfolded protein burden without further stressing their protein degradation machinery by expelling protein aggregates into the extracellular space. Misfolded protein secretion is emerging as a common clearance mechanism in a variety of cell types and contexts,[111–115] and may be especially important in cells that need to persist for decades at a time, such as cardiomyocytes and neurons.[116,117]

2.4 Extended modules for eukaryotic proteostasis

Whereas the core modules of the PN (synthesis, folding, and degradation) are largely conserved throughout all cellular life, the increase in complexity of the eukaryotic cell creates additional challenges that require the extension of tasks that fall under the remit of the PN. This includes the requirement to traffic proteins across membranes into other subcellular compartments (bacteria have a single cytoplasm-like intracellular compartment) and to assemble proteins translated from spatially distant genes into the same complex (bacteria have a single chromosome, mainly possess homomeric protein complexes, and generally group subunits of the same heteromeric complex together as operons).[118–120]

2.4.1 Targeting

According to the Human Protein Atlas, one-quarter to one-third of all human protein-coding genes may be localized in the cytoplasm.[121] Most cytoplasmic proteins are also present in at least one additional subcellular compartment.[121,122] Therefore, most newly synthesized proteins need to be trafficked from cytoplasmic ribosomes to other locations.[119] This targeting can occur cotranslationally or posttranslationally, with an array of targeting factors responsible for recognizing one or more targeting signals within the protein's sequence and guiding it to the specified subcellular compartment.[123] For cases in which targeting fails, general and specialized PQC factors detect such mislocalized proteins and mediate triage to pro-targeting or prodegradation pathways, much like in misfolded protein triage.[118,124] The machinery and mechanisms for PQC of mislocalized proteins are no less complex than those for misfolded proteins, which suggests that these targeting and PQC factors constitute an important module of the PN.[125]

The targeting burden could be eased even before protein synthesis takes place, by transporting mRNAs and/or ribosomes so that they are poised for translating specific protein classes at the most relevant subcellular locations. Local targeting of the translation machinery is especially important in neurons, where random ribosomal translation would raise situations in which nascent proteins need to be transported large distances to the correct location (e.g., from the cell body to distant synapses) without aggregating or interfering with other aggregation-prone cytoplasmic proteins.[126] Local translation could be further aided by intracellular ribosome heterogeneity—the finding that cells contain distinct ribosome sub-populations that display preferences for different mRNAs.[127]

2.4.2 Protein complex assembly

In a manner similar to that of proteins needing transport to subcellular compartments, mechanisms exist to target nascent chains cotranslationally or posttranslationally to their target protein complexes. Cotranslational as-sembly is more prevalent in eukaryotes than previously thought,[120,128] and it could be the predominant method for assembling obligate protein com-plexes that do not have specialized assembly factors.[129] It is well-established that proteins functioning exclusively as subunits of protein complexes are targeted for degradation when they are orphaned (i.e., without their obligate binding partners at stoichiometric ratios).[130−132] Examples of such obligate complexes include several components of the PN, such as the ribosome,[133] molecular chaperones TRiC/CCT and prefoldin,[134] and both 19 and 20S proteasome.[135] The generation and maintenance of stoichiometric protein complexes are major functions of the PN.[136]

Clues to the emphasis placed on PQC at the protein complex level might be provided by studies characterizing proteomic perturbations in cells with additional copies of one or more chromosomes (e.g., Down syndrome patients, who have an extra copy of chromosome 21).[137−139] When measuring levels and turnover rates for proteins encoded on extra chro-mosomes, proteins from heteromeric complexes are targeted to degradation or cytoprotective sequestration systems much more effectively than those functioning alone or as homomers. Therefore, the PN buffers against extra gene dosage effects of aneuploidy primarily by recognizing excess subunits and targeting them to clearance systems.[140]

2.5 Executive modules to regulate the proteostasis network

Components that collectively constitute the PN are broadly categorized into distinct modules based on their predominant role within the network. However, as we have discussed, several components have important functions across modules. For example, the ribosome directs the translating polypeptide through the folding landscape, whereas molecular chaperones decide whether a misfolded protein needs to be degraded and which clearance strategy is employed.

Borrowing from systems biology terminology, the network consists of nodes (proteins that make up the network) linked by edges (pairwise genetic, physical, or functional interactions between proteins in the network). As a rule of thumb, there will be more edges linking nodes within a module than between modules. However, a degree of interconnection between modules is crucial for maintaining a robust proteome, so that regardless of where the initial stress to the system takes place, the entire network can adapt to mount a coordinated response that restores proteostatic balance.

2.5.1 Stress responses remodel the proteostasis network to restore proteostasis

In light of how close much of the proteome is to its stability limits, even subtle changes in environmental conditions could risk widespread aggregation if not buffered appropriately. Although molecular chaperone levels (especially holdase functions) have evolved to provide a degree of buffering capacity around the physiologic limits of the organism's operating conditions, certain stresses can exceed this capacity. Therefore, cells can activate a variety of stress responses that remodel the PN in a manner best suited to protect the proteome from dangers associated with that stressor.

The activation of each stress response is controlled by one or more transcription factors. The most extensively studied stress response is the heat shock response (HSR) mediated by heat shock factors (e.g., HSF1 and HSF2)—although it is activated by a wide range of proteotoxic stressors in addition to acute temperature elevation.[141] A similar assortment of stressors can also activate the integrated stress response (ISR) via kinases of the eukaryotic translation initiation factor eIF2α.[142] Proteotoxic stresses in organelles trigger specialized unfolded protein responses, such as in the ER (unfolded protein response [UPR]^{ER})[143] or mitochondria (UPR^{mt}).[144] The UPR also activates the ISR, and involves some of the same machinery. Reactive oxygen species (ROS) activate the oxidative stress response via NRF2[145]; low oxygen levels activate the hypoxic stress response via hypoxia-inducible factors (HIFs).[146]

The mechanisms through which stress responses are activated at the transcriptional level revolve around sensing the free capacity of one or more PN machineries.[147] For example, HSF1 activation is thought to rely on a simple chaperone titration model, in which its binding to HSP70 and/or HSP90 under physiologic conditions keeps HSF1 inactive. Proteotoxic stresses titrate away these two major chaperone hubs of misfolded protein triage, releasing HSF1 to translocate to the nucleus and activate transcription of its target genes.[148] A similar chaperone titration mechanism may exist for the ER HSP70 isoform BiP with the IRE1 arm of the unfolded protein response,[149] although this model remains controversial.[150] Other stress responses work (e.g., oxidative or hypoxic stress response) with the same concept but instead keep their central transcription factor from binding target genes under physiologic conditions by constitutive ubiquitin–mediated proteasome degradation by their cognate E3 ubiquitin ligase. Oxidative or hypoxic stresses result in stabilization of NRF2 or HIF1α, respectively, owing to reduced ubiquitylation by their cognate E3 ubiquitin ligase, allowing activation of their gene expression programs under those stresses.

In general, most proteotoxic stress responses involve boosting expression of cytoprotective factors that help contain the misfolded protein load (e.g., HSPs during the HSR) while reducing expression of all nonessential proteins. Decreasing the rate of protein synthesis is a conserved mechanism to maintain nascent proteome quality and reduce PN burden in a range of programmed events during development and physiologic homeostasis. Proteotoxic stress can trigger translation arrest (via elongation pausing) in a manner similar to that of chaperone-titration models described earlier for transcription factor regulation (e.g., owing to sequestration of the de novo folding CLIPS network by misfolded proteins).[151] In other cases, the translation machinery is targeted directly, as is achieved by eIF2α phosphorylation to inhibit global cap-dependent translation during the ISR.[142] The ISR also leads to the specific translation of stress-protective proteins via upstream open reading frames present in their mRNAs and can ultimately trigger cell death via apoptosis if the stress fails to be successfully resolved. Finally, in addition to changing the synthesis and folding modules, stress responses can boost expression of clearance machineries, as happens with two of the three arms of the UPR^{ER} (IRE1–XBP1 for ER-associated degradation via the ubiquitin–proteasome system; PERK-ATF4 for autophagy).[143]

2.5.2 Organism-wide proteostasis control

Much of what we understand of proteostasis at the molecular level stems from studies in a cell-autonomous setting (i.e., proteostasis strategies employed by the cells directly subjected to the specific proteotoxic stress). Such proteostasis strategies, which likely evolved as an adaptation at an individual cell level, have the potential to act as a double-edged sword in the context of multicellular life. For example, protein clearance via exocytosis (discussed earlier) was proposed as a mechanism for the spread of disease-associated misfolded proteins from neuron to neuron in neurodegenerative diseases.[152] More generally, cytoprotective responses that evolved to promote cell survival can have deleterious consequences for a multicellular organism (e.g., cancer growth; cellular senescence-associated inflammation). Therefore, understanding how proteostasis is coordinated and maintained at different scales, especially at the whole-tissue and whole-organism levels, is an important next step in the field; great strides have been made (mainly using the model organism *Caenorhabditis elegans*).[33,153–156]

3. Proteostasis network dysregulation in aging

3.1 Proteostasis network size as determinant of species life span

Important insights into the role of the PN in aging can be gained from comparing species with differing life spans, although these need to be confirmed for relevance in human longevity: in addition to possessing mechanisms of longevity conserved throughout eukaryotic life, most species have evolved their own unique prolongevity adaptations according to their specific niche.[157] Nevertheless, proteostasis capacity is one of the major predictors of interspecies differences in life span (as is insulin/insulin-like growth factor-1 signaling [IIS], which also has a marked impact on proteostasis).[157] For example, the fidelity of protein synthesis correlates with the maximum life span across 17 rodent species.[158] A broader computational analysis of the PN across 216 metazoans found a positive correlation between the size of an organism's chaperome and its longevity.[159] Because PNs are proposed to drive diversification of the proteome by enabling evolution of novel protein functions,[160] it may be possible to draw a direct line (albeit a correlative one) between an organism's PN size, its proteome complexity, and its maximum life span.

In parallel with the increase in PN size, the expression and activity levels of both folding and degradation modules within the network correlate positively with longevity. One comparative study of the PN in skin fibroblasts from phylogenetically similar longer-lived versus shorter-lived mammals found higher levels of HSPs in the longer-lived partner, as well as generally higher proteasome and macroautophagy activities.[161] There were phylogeny-specific differences: most notably, longer-lived rodents had elevated ATP-dependent machineries (HSP70 and HSP90 levels as well as 20S proteasome activity), whereas longer-lived bats had elevated levels of the major ATP-independent sHSP HSP27. Longer-lived marsupials appeared to have elevated levels of all tested proteostasis machineries. This study suggests that longer-lived mammals have evolved higher proteostasis capacities than their shorter-lived counterparts, although the strategies through which they achieve this varies among species.

A more recent interspecies comparison of fibroblasts isolated from a broader range of animal species with different life spans found a negative correlation between life span and proteome turnover rates (at least for highly abundant proteins).[162] Importantly, a direct comparison of fibroblasts from the long-lived naked mole-rat versus the shorter-lived mouse found that naked mole-rat fibroblasts were more resistant to nascent proteotoxic stress—but only when the comparison was made in the quiescent state. The authors suggest that longer-lived species evolved more energetically efficient mechanisms for PQC rather than simply relying on high basal turnover rates to clear damaged proteins stochastically, which would be energetically demanding and increase ROS production. Although this comparison could indicate more general proteostasis trends between shorter- versus longer-lived organisms, naked mole-rats appear to be special cases with regard to many of their cellular and molecular stress adaptations[163] and are the only mammals identified that do not appear to follow Gompertzian mortality laws.[164] It will be important to establish whether the increased proteotoxic stress resistance holds true for other longer-lived mammals.

3.2 Proteostasis network remodeling with age

Interspecies comparisons provide clues to evolutionary requirements and constraints for longer life spans. However, the vast majority of research linking proteostasis with aging comes from direct comparisons of the same species or cells at different stages of life. Human studies have been necessarily correlative (although important insights have emerged from large population-level datasets). Here, we focus on changes in the PN gleaned from aging model organisms—most commonly budding yeast, nematode worms, fruit flies, and mice—and, where possible, describe whether the same mechanisms have been observed in human cells or studies.[165] As mentioned in the previous section, each of these species likely contains niche-specific adaptations not conserved in humans. However, because the modules of the PN are highly conserved among all eukaryotes, it is highly probable that the general mechanisms that change with age are conserved, even if the specific proteins or machines involved differ among species. A higher degree of confidence can also be attained if the same observation is made for two or more model organisms from distant phylogenies. One multiple system-level analysis identified all three core PN modules, as well as protein transport/targeting and complex assembly, as being significantly altered during normal aging across worms, fruit flies, mice, and humans.[166]

3.2.1 Dysregulation of core proteostasis network modules

The multiple systems-level analysis referenced earlier[166] included transcriptomics data from the large-scale human genome-wide association studies of two postmitotic tissues: skeletal muscle and the hippocampus. In both tissues, expression of genes associated with synthesis and degradation (both ubiquitin–proteasome system and macroautophagy) declined with age and agreed with a more recent analysis linking reduced expression of the protein translation machinery with human longevity across a variety of tissues (including skeletal muscle).[167] These changes correlate well with aging-associated changes seen in other organisms—not only in the study in question (e.g., in mouse skeletal muscle and hippocampi) but also across multiple studies in a range of model organisms, summarized in this section (Fig. 4.3).

It would be easy to assume that aging-associated loss of proteostasis is caused by a decrease in levels of one or more of its constituent nodes or modules. However, the loss in this case refers to the loss of balance in physiologic protein homeostasis, rather than necessarily a decline in, for instance, chaperoning capacity (an increase in which is considered to enable malignancy[168,169]). It might therefore be more useful to conceptualize this hallmark as proteostasis dysregulation, or imbalance—regardless of the direction of the imbalance.

Multiple reasons have been proposed for why proteostasis is dysregulated with age—generally tied to whichever theory of aging is prominent at the

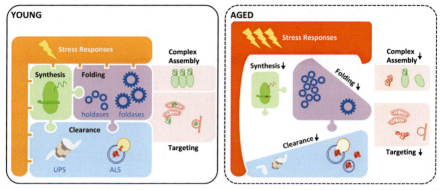

Figure 4.3 *Dysregulation of proteostasis network (PN) module composition and integration with age.* In young cells, the modules of the PN are well-balanced. As cells age, the modules are dysregulated. Fidelity of protein translation declines, which may be compensated for by the cell reducing its synthesis rates. The folding module balance shifts from ATP-dependent foldases (e.g., heat shock protein [HSP]70–HSP90 system, TRiC/CCT) to increasing reliance on ATP-independent holdases (e.g., small HSPs), potentially reflecting reduced energy consumption with age. Similarly, ubiquitin–proteasome system (UPS) expression and activity declines, and although the autophagy–lysosome system (ALS) may appear to increase, several components of this system show functional decline (e.g., accumulation of autophagosomes and aggregates; lysosomal deacidification and leakage). The extended PN modules also lose fidelity in maintaining protein complex stoichiometry and protein subcellular compartmentalization. Finally, the stress responses lose plasticity owing to rigid adaptation to chronic stress levels in aged cells and tissue. All of these changes result in a weakening in overall communication between PN modules and thus reduce proteome robustness.

time. For example, one explanation revolves around the energetically expensive nature of protein turnover (synthesis, folding, and degradation), which necessitates a high demand for ATP synthesis. ATP synthesis via the mitochondrial electron transport chain leads to ROS production, which involves free radicals that cause irreversible damage to macromolecules. Because irreversibly oxidized protein adducts (e.g., carbonylated proteins) accumulate with age and need to be cleared by the proteasome, reducing proteome turnover rates would therefore serve to slow ROS-induced protein oxidation owing to lower ATP demand.[101,170] Understandably, the popularity of this hypothesis is tied to the high regard for the free radical theory of aging, which has waned in recent years.[171] The fundamental reasons are likely multifactorial and may be linked in some way to differing selection pressures for proteome fidelity after an organism's reproductive potential has passed.[172,173] In agreement with this hypothesis, most genetic PN manipulations in *C. elegans* that increase life span reduce reproductive fecundity, which suggests the existence of an evolutionary trade-off between proteostasis resource allocation to the soma versus the germ line.[101,174]

3.2.1.1 Decline in protein synthesis

Proteome turnover is dictated by relative activities of the synthesis and degradation modules—both of which decline with age.[175] Therefore, it is difficult to disentangle cause from effect: that is, are synthesis rates declining because proteins are not being degraded quickly enough (which would saturate the available chaperone machinery), or are degradation rates declining because there are not as many proteins being synthesized in older organisms? Clues can be provided by studies that measure age-associated changes in error rates during protein synthesis. These generally find that, in addition to declining global protein synthesis rates, translational fidelity decreases with age.[176,177] A paper extended this finding to show that (1) ribosomes stall more often at regions that are challenging to synthesize (e.g., polybasic amino acid sequences) in aged yeast and worms, and (2) these stalled ribosomes and nascent chains are cleared less effectively with age.[178] This work may explain why more nascent chains, ribosomes, and CLIPS machineries are found in aggregates in aged cells.[178–180] Reducing the rate of protein translation may therefore improve the chances for ribosomes to recover from stalling and/or increase the triage time available for RQC-linked machineries to degrade the stalled complexes. Even without factoring in a reduction in translational fidelity, the general decline in downstream PN modules with age (e.g., folding, degradation, stress responses) might necessitate a reduction in the rate of protein synthesis to allow these downstream modules to focus on conformational maintenance of the existing proteome rather than strain to fold nascent chains.[181] The finding that ribosome biogenesis and protein synthesis machineries are downregulated at the transcriptional level in aged worms[182,183] and mice[184] supports the hypothesis that downregulation of the synthesis module is an active, programmed response by the cell rather than simply another symptom of proteostasis machineries breaking down with age.

3.2.1.2 Remodeling of chaperone-mediated protein quality control

Early experiments measuring changes in chaperone levels with age reported varying trends, with an increase, decrease, or no change in old versus young states.[185] This appeared to be at odds with remarkably consistent data across a variety of organisms, cell types, and aging models, showing that induction of

HSP expression in response to stress was considerably dampened in older cells.[186] These discrepancies can be reconciled by the fact that aged cells display chronically elevated stress responses,[187] including the HSR (discussed later).[188,189] Therefore, higher steady-state levels of HSPs measured in old cells were presumably due to being compared with unstressed young cells.

A landmark 2014 study in the aging—proteostasis field further clarified several aspects of chaperome remodeling with age.[190] After defining the human chaperone complement as 336 proteins, the authors compared the expression levels and functional interaction network for this chaperome with healthy aging and neurodegeneration. They found that roughly half of the chaperome was significantly dysregulated during healthy brain aging ($\sim$30% down and $\sim$20% up). The aging-repressed chaperome consisted mainly of ATP-dependent cytoplasmic machineries (e.g., HSP40-HSP70-HSP90 and TRiC/CCT), whereas the induced set contained mostly ATP-independent sHSPs (although ATP-independent machineries were also present in the age-repressed set). The same chaperome subsets were dysregulated more dramatically in brains of Alzheimer's, Parkinson's, and Huntington's disease patients, which suggests a threshold phenomenon separating healthy from pathologic aging, rather than being completely unrelated processes.

This study had multiple implications. For one, the aging-induced ATP-independent chaperone subset dovetails neatly with earlier work showing elevated sHSP levels in longer-lived worms[191] and fruit flies.[192] The reduction in ATP-dependent chaperones (foldases) and increase in ATP-independent chaperones (holdases) may have evolved to maintain proteostasis under lower ATP levels with age, as discussed earlier. This foldase-to-holdase switch could indicate a more global shift in PQC strategy with age (i.e., from minimizing the misfolded protein load of the cell or organism in total by refolding or degradation to limiting the damage the misfolded protein load can trigger by sequestering it away from the functional cellular milieu) (Fig. 4.2C). The proteome quality versus proteome damage limitation balance mediated by the PN appears consistent when comparing other cell states characterized by vastly different cell proliferation rates (e.g., neural stem cells versus differentiated neural lineages,[193] proliferating versus quiescent dermal fibroblasts,[110] and cancers versus neurodegenerative diseases).[168]

Another finding from that study, which has since been formalized by various interaction-based approaches, is that the connectivity between members of the chaperome is dysregulated with aging and neurodegeneration, not just expression levels. A key feature of the chaperome connectivity network, sometimes referred to as the epichaperome, is its ability to be dynamically regulated in young and healthy cellular states so that it can adapt to different types of stressors and misfolded protein loads.[194] Chronic pathologic states, by contrast, lock the epichaperome into more rigid conformations owing to the constitutive requirement for chaperoning specific dysregulated proteins or processes (e.g., mutated or overexpressed oncoproteins; amyloidogenic oligomerization and aggregation).[195] Therefore, upregulated and more connected networks do not necessarily indicate healthy proteostasis and increased stress resilience if they are less adaptable to additional or different stresses. The epichaperome remodeling concept could help explain the aging- and neurodegeneration-induced upregulation of sHSPs (known to form higher-order oligomeric complexes) as well as earlier findings of high—molecular weight HSP90 multichaperone complexes in cancers.[196,197]

3.2.1.3 Shifting degradation and sequestration machineries

Almost all degradation systems have been reported to decline with age (Fig. 4.4). For example, decreased expression of proteasomal subunits, changes in 20 and 19S subunit composition, disassembly of the 26S holo-complex, and reduced available proteasome activity (e.g., owing to titration by age-accumulated oxidized or aggregated proteins) all likely contribute to aging-associated proteasome dysfunction.[198,199] Similarly, decreased flux through macroautophagy (both bulk and selective) and chaperone-mediated autophagy appears to be conserved across a range of organisms.[10,200] However, there is a degree of variation when comparing different tissues. Some tissues show less pronounced differences, or even upregulation, with aging.[199] Although there is likely some degree of tissue-specific variation in proteostasis dysregulation with age, part of this discrepancy could also be related to cross-talk built into the PN. Degradation systems are no exception, and upregulation of alternative clearance pathways such as autophagy and sequestration are a common cellular response to acute proteasome inhibition or depletion.[67,199,201] Although it is not clear how conserved this proteasome-to-autophagy switch is with more chronic proteasome impairment (as would be the case during aging), it was reported in one study comparing young versus old mouse brains, as well as in vitro replicatively senescent fibroblast lines.[202] Switching to autophagy might serve an analogous function to the foldase-to-holdase switch, shifting from the highly specific and ATP-intensive series of steps leading to substrate ubiquitylation and proteasomal degradation one at a

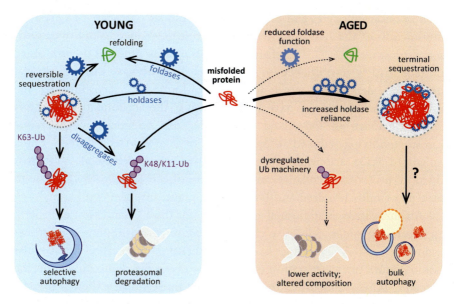

Figure 4.4 *Misfolded protein quality control strategies change with age.* In young cells, the balance of proteostasis tips toward ATP-dependent refolding or selective degradation of a misfolded protein. Misfolded proteins may be temporarily and reversibly sequestered in Q-bodies or juxtanuclear quality control compartment/aggresomes. In older cells, the decline in ATP levels and dysregulation of active protein quality control machineries (especially the ubiquitin–proteasome system) tips the balance toward terminal sequestration by holdases. The extent to which sequestered cargoes are turned over by bulk macroautophagy is unclear.

time toward a more promiscuous and bulk recycling pathway that can, in one step, reduce the misfolded protein load and replenish the nutrient supply. The switch to autophagy also need not be predicated on having a defective ubiquitin–proteasome system, as highlighted by a systemwide analysis of ubiquitylation patterns in worms.[203] Rather than being merely a readout of decaying proteostasis, the authors found that elevated DUB activity in aged worms could account (at least partially) for the global reduction in ubiquitylation they observed, which suggests that a programmed response was at play during aging.

Regardless of how programmed this shift is in degradation strategy, it appears to be a temporary or only partially successful measure: eventually, protein aggregates will accumulate in older cells. This may be due to the gradual accumulation of degradation-refractory material over the cell's life span: for example, lipofuscin, a highly cross-linked oxidized macromolecular mixture that appears in lysosomes and cytoplasm of long-lived cells such as cardiomyocytes and neurons.[204] It could also be related to eventual dysfunction of the lysosome itself with age (e.g., reduced activity of lysosomal proteases).[205,206] Therefore, regardless of how much of the misfolded protein load is re-trafficked to the autophagy pathway, the lysosome is unable to degrade it proteolytically. Furthermore, according to the mitochondrial—lysosomal axis theory of aging, the accumulation of intracellular garbage in long-lived postmitotic cells triggers a degenerative cascade involving reduced autophagic turnover of damaged mitochondria, enhanced ROS accumulation and oxidative stress, decreased ATP production, proteostasis collapse, and activation of cell death pathways due to leakage of mitochondrial and lysosomal contents into the cytoplasm.[207]

An alternative or additional explanation for the accumulation of protein aggregates in older cells is a change in biophysical properties of stress-induced foci. A growing body of literature implicates intracellular liquid—liquid phase separation (LLPS)—the reversible demixing of specific proteins, RNA, and other molecules from the cellular milieu into biomolecular condensates of defined composition and function—as a crucial mechanism for a variety of developmental and physiological functions.[208,209] Examples of such membraneless compartments include several stress response and PQC-associated structures, including stress granules, p62-bodies, and the nucleolus, and are likely to expand to other reversible spatial sequestration sites (JUNQ/aggresome and Q-bodies).[67] Changes to LLPS properties that make biomolecular condensates less liquid-like (a liquid-to-gel transition) are proposed to contribute to neurodegenerative disease pathology, potentially because disease-associated amyloidogenic proteins form stable oligomers or fibrils that interfere with condensate fluidity and reversibility.[210] In agreement with this hypothesis, artificially targeting IPOD substrates (e.g., prions or polyglutamine-repeat proteins) to the JUNQ impairs the dynamics of the whole compartment and eventually leads to cell death.[99,211,212] Although not explicitly shown to occur in the context of healthy (nonpathologic) aging, it would be unsurprising if liquid-to-gel like transitions become more common with age.

3.2.2 Dysregulation of extended and executive modules
3.2.2.1 Protein targeting defects
Spatial compartmentalization is a key hallmark of health from the subcellular to the organismal level, maintaining the integrity of barriers and preventing local perturbations from spreading throughout the system.[12] Subcellular partitioning into organelles—each with its specific proteome—is a dynamic

rather than static process, with transport between compartments an important part of homeostatic signaling circuits (e.g., nuclear translocation of cytoplasmic transcription factors during stress response). The subcellular localization of the protein also affects systems available for its triage (e.g., different degradation circuits exist for the same misfolded protein in the nucleus versus cytoplasm[78]).[65] Because of the importance of compartmentalization in these major aspects of proteostasis, disintegration of these barriers is considered a main cause of aging-associated frailty and susceptibility to related pathologies.[12] For example, nuclear—cytoplasmic transport defects are a common mechanism in the pathology of several neurodegenerative diseases and are attributed to some combination of nuclear pores becoming more leaky with age, and a direct consequence of disease-causing proteins (e.g., FUS, TDP-43) forming juxtanuclear aggregates that clog the pores.[213] Disturbances in communication are similarly problematic between the nucleus and mitochondria, especially for nuclear-encoded transcripts that need to be translated and imported into the mitochondria.[214] Furthermore, there is growing evidence that local translation is dysregulated in neurodegeneration and other nervous system pathologies through multiple mechanisms (e.g., through mutations in RNA-binding proteins that regulate mRNA transport along neurites).[215]

3.2.2.2 Loss of protein complex stoichiometry

Global loss of protein complex stoichiometry has been observed in several aged model organisms.[180,183,216] Because of the importance of protein complexes for almost all cellular processes, it is unsurprising that protein complex misassembly is a major driver of human disease.[217] Long-lived protein complexes (with a turnover rate of months or even years) may be especially important in aging-associated dysregulation, because these include central gatekeepers of cellular homeostasis and longevity (e.g., nuclear pore complexes, nucleosomes, and supercomplexes of the mitochondrial electron transport chain).[218,219] Their decline with age, despite being transcribed at similar levels, may be related to a decline in general PQC mechanisms and machineries described earlier, or a decline in levels of specific assembly factors, as is the case of nuclear pore complexes.[220,221] Dysregulation of protein complex homeostasis also manifests in other aging hallmarks, such as mitochondrial dysfunction, which is at least partly related to a decline in mitochondrial electron transport chain supercomplexes.[222]

Loss of protein complex stoichiometry is not problematic just because it creates extra substrates for PQC machineries to process. Protein complexes of the PN are themselves dysregulated in aging and several pathologies, as described earlier. Most notably, substoichiometric components of the protein translation machinery (e.g., ribosomes and translation factors) accumulate in yeast, worms, and turquoise killifish models of aging,[180,183,216] although conflicting reports exist regarding whether ribosome stoichiometry is lost with age in mice.[180,223] Accumulation is linked to a decoupling of transcription with translation (i.e., mRNA levels correlated less well with protein levels with age), which also correlates with a decline in proteasome activity despite an increase in expression.[180,183]

3.2.2.3 Decline in stress resilience

The importance of stress responses in aging-associated frailty was discussed in Section 3.2.1.2. However, it is highlighted by findings that the life span shortens nonlinearly with increasing stress severity.[181,224−226] Therefore, although induction of stress responses is an evolutionarily conserved feature of aging,[187] it must be taken hand–in–hand with a decline in biological resilience.[227] These two seemingly counterintuitive findings can perhaps be reconciled by the existence of stress attenuation feedback loops activated with prolonged or chronic stress[189] and/or the desensitization of chronically activated stress response pathways to additional stressors.[224,226,227]

In addition to the stress responses mentioned earlier (e.g., HSR, oxidative stress response, integrated stress response, unfolded protein responses), all of which decline with age, another stress–responsive signaling network needs to be discussed in any conversation on aging and proteostasis. This network is the major link between nutrient sensing and homeostatic responses at a cellular level and includes the key metabolic hubs of IIS, the protein kinases mTOR and AMPK, and a class of NAD^+-dependent deacetylases known as sirtuins.[228] As one might expect, all four of these hubs are highly interconnected to balance proteostasis according to the cell's metabolic state, and are thought to collapse around activation or inhibition of the FOXO family of transcription factors.[228] Each of these proteins is the subject of a vast body of literature on aging, proteostasis, and signaling communities.[229] In the simplest terms, IIS and mTOR activation lead to an increase in synthesis and growth, whereas AMPK and sirtuins arrest synthesis and growth and instead promote recycling of cellular material. This system feeds into the PN through several layers. For example, although much of the research on AMPK and sirtuin–mediated degradation has focused on starvation-based activation of bulk macroautophagy, other work suggests a more sophisticated mechanism relying on the E3 ubiquitin ligase CHIP.[230] Under unstressed conditions, CHIP regulates insulin receptor turnover through the endosome—lysosome system. Under conditions of stress, it is sequestered by the increased misfolded protein burden to HSP70 (and/or HSP90) complexes, reducing insulin receptor turnover and increasing IIS activity.[231] Because IIS activity is negatively correlated with life span across multiple organisms and aging models,[232] this mechanism would provide an elegant link between high stress levels and reduced life span.

4. Interplay with other aging hallmarks

According to the framework of Lopez-Otin et al., loss of proteostasis is one of the four primary hallmarks of aging: it causes damage that triggers the antagonistic hallmarks (responses to damage), which in turn lead to eventual onset of the integrative hallmarks (culprits of the phenotype)[2] (Fig. 4.5). The other three primary hallmarks—genomic instability, telomere attrition, and epigenetic alterations—center on chromosome-level alterations with age. According to the central dogma, these alterations will translate through to the protein level. We have discussed how aneuploidy in diseases such as cancer or Down syndrome creates additional stress for the PN, especially for nonstoichiometry protein complex subunits.[139,140] To take another example, aged mouse and human tissues display increased expression of genes lacking CpG islands, a large proportion of which encode secreted proinflammatory proteins.[233] In addition to direct consequences such as acquisition of the senescence-associated secretory phenotype, this increased expression is likely to represent a chronic load on the secretory components of the PN—which, as discussed earlier, is a major source of proteostasis dysregulation via ER stress and the unfolded protein response.

We have also described the links between proteostasis dysregulation and two of the antagonistic hallmarks of aging: deregulated nutrient sensing and mitochondrial dysfunction. Metabolic rewiring has profound effects on proteostasis through changing the relative balance among IIS, mTOR, AMPK, and sirtuin activation.[228] The reverse is also true: proteostasis dysregulation feeds back into nutrient sensing pathways because of the high energetic demands of proteome turnover and QC. A similar bidirectional relationship exists between mitochondrial and proteostasis dysfunction.[242] Aside from proteostasis of the mitochondrial proteome (a specialized network in its own right, with extensive links to aging[255]), mitochondria also act as regulators of cytoplasmic and ER proteostasis, such as by importing cytoplasmic aggregation-prone proteins for degradation via mitophagy.[242]

84 Molecular, Cellular, and Metabolic Fundamentals of Human Aging

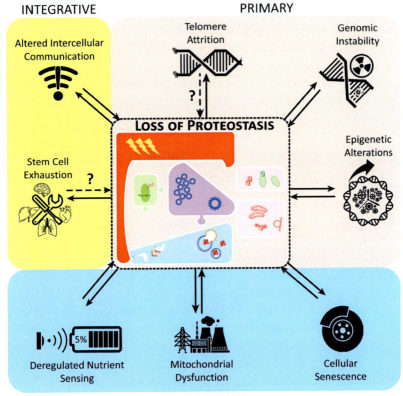

Figure 4.5 *Loss of proteostasis as a central process linking other hallmarks of aging.* As a primary hallmark of aging (i.e., causative of damage), loss of proteostasis has profound effects on antagonistic and integrative hallmarks. It is even able to influence, and be influenced by, other primary hallmarks, possibly resulting in detrimental feedback loops. *Dashed arrows* and question marks indicate minimal or missing evidence for that direction of interconnection (i.e., whether telomere attrition or stem cell exhaustion can trigger further proteostasis dysregulation). Table 4.1 provides recommended reading for each of these interconnections.

Under conditions of extreme stress or other situations with a high misfolded protein burden, mitochondria can trigger apoptotic signaling pathways.[256] Although there is a lack of comprehensive literature covering the relationship between proteostasis and the third antagonistic hallmark, cellular senescence, clear conceptual links are backed up by studies showing altered PN composition and activity in senescent cells,[202,248,257–259] as well as PN modulation as a trigger for inducing senescence.[260,261]

Finally, stem cell exhaustion and altered intercellular communication—the integrative hallmarks—could be seen as direct consequences of proteostasis

Table 4.1 Recommended reading for links between loss of proteostasis and other hallmarks of aging. Primary hallmarks are in red, antagonistic hallmarks are in blue, and integrative hallmarks are in yellow.

Hallmark of aging	Recommended reading
Epigenetic alterations	233–236
Telomere attrition	237–239
Genomic instability	240,241
Mitochondrial dysfunction	242–244
Deregulated nutrient sensing	245,246,228
Cellular senescence	247–249
Stem cell exhaustion	250–252
Altered intercellular communication	153,253,254

dysregulation. Not only is there extensive evidence that the elevated PN capacity in stem cells is fundamentally important for maintaining their identity, experimental manipulation of PN machineries in stem cells to make them more like their differentiated counterparts is often sufficient to initiate differentiation, thus establishing a rare causal link between the two.[262,263] By contrast, how proteotoxic stress in one cell is communicated to surrounding cells and/or throughout the organism is a relatively recent field (see Section 2.5.2). Nevertheless, although it is in its infancy, the cell nonautonomous proteostasis field is revealing this additional layer of stress regulation and communication that is likely to prove crucial for maintaining tissue and organism health during aging.[264,265]

5. Targeting proteostasis for healthy aging

The pursuit of interventions that extend the healthy human life span is one of the next frontiers in therapeutics. The difficulty stems not only from the fact that aging is generally considered a degenerative multifactorial wear-and-tear process, but that interventions to treat people who have no specific pathology must necessarily meet a higher standard, demonstrate efficacy, and lack toxicity. Perhaps the progress of the Targeting Aging with Metformin study, the first US Food and Drug Administration–approved clinical trial with aging as its primary end point,[266] will pave the way for more health span–modulating interventions in the future. An important qualifier here is that metformin has been in clinical use for over 60 years for type 2 diabetes and has a comprehensive and robust safety profile;

demonstrating the same confidence in a novel, nonapproved intervention is likely to provide additional challenges.

Metformin's life extension properties can be attributed to attenuation of almost all hallmarks of aging to varying extents,[266] yet a central component is via AMPK activation—which, as discussed earlier, leads to a reduction in protein synthesis and activation of autophagy. Many of these effects are mediated through AMPK's inhibition of mTOR signaling, and indeed the direct mTOR inhibitor rapamycin and related rapalogs are also promising candidates for human longevity.[267] In fact, most aging interventions in development center on the IIS—AMPK—mTOR—sirtuin signaling axis, which is unsurprising given the long-standing literature linking caloric restriction to longevity.[268] That reducing protein synthesis or activating autophagy through other means also has life span—promoting effects[10,176] suggests that these two processes are drivers of longevity rather than merely bystanders to some other IIS—AMPK—mTOR—sirtuin-related effects. For example, modulating either the integrated stress response or unfolded protein responses—both of which reduce protein synthesis and activate degradation—increases health span and life span.[264,269] Increasing the fidelity of protein translation also extends both health span and life span,[176] with one study reporting additivity between a hyperaccurate ribosome mutant and rapamycin treatment in fission yeast, worms, and fruit flies.[270]

That is not to say that all longevity-targeted PN interventions revolve around the synthesis and autophagy systems. Modulation of virtually any PN module has reported effects on life span and/or health span.[197] Genetic screens in model organisms reveal that increasing chaperone or proteasome levels has a positive effect on longevity.[181] Even with regard to long-lived mutants of the insulin signaling pathway, mechanisms other than synthesis and autophagy exist for different tissues, such as increased mitochondrial biogenesis and respiration in the fat body and higher proteasome assembly and activity in the gut of fruit flies.[271] Mitochondrial dysfunction appears to be an important factor in proteostasis-related interventions, as demonstrated by a study showing that impairing electron transport chain function early in life can protect against aging-associated proteostasis collapse by inducing expression of cytoprotective sHSPs in a mitochondrial HSF1-dependent manner.[272] Similarly, proteostasis modulators have been identified as hits in screens targeting other aging hallmarks, such as senotherapeutics.[273–275]

PN-targeting therapeutics clinically developed as anticancer agents can also provide clues as to how they could be repurposed for healthy aging interventions. For example, proteasome inhibitors have been approved for multiple myeloma since 2003,[276] and HSP90 inhibitors have seen multiple waves of therapeutic interest for solid and hematologic malignancies.[277] Although the goal of these agents at a cellular level is to kill cancer cells selectively (the opposite of what is desired for longevity in healthy cells), the cytoprotective responses and adaptations of the tumors that limited the efficacy of some of these agents could be desirable for aging interventions. A main reason for the halted development of early HSP90 inhibitors was induction of the HSR via HSF1 activation, thus promoting cancer cell survival.[277] Low-dose HSP90 inhibition may therefore be cytoprotective for healthy aging via HSF1 induction, as supported by an unbiased screen for transcriptomic geroprotectors.[278] This cytoprotective mechanism feeds into the hormesis concept (i.e., the reported beneficial effects of mild stress, including life span extension).[279]

6. Concluding remarks

Whereas a large body of research has focused on components of the PN that are modified in age-associated pathologies (the most prominent being neurodegenerative diseases and cancers), until recently, our understanding of how these change in normal aging has been much more limited. Among the broad spectrum of proteostasis dysregulation and remodeling processes that occur with age, the most conserved changes include a decline in protein translation fidelity, a shift in PQC strategy from ATP-dependent refolding or proteasomal degradation to ATP-independent sequestration and/or autophagy upregulation for recycling of resources, a loss of protein complex stoichiometry and interorganellar trafficking, and a decline in robustness and resilience in stress responses. Modulations that restore the most well-characterized of these changes—namely reducing protein synthesis and boosting autophagy—are among the most promising health span interventions in clinical testing. However, that numerous other genetic and chemical perturbations of core, extended, and executive PN modules affect longevity suggests a much wider array of targets and intervention strategies that remain untapped. A progressive shift in emphasis within the proteostasis communities toward assessing global changes at the level of the entire PN (rather than individual machineries in isolation) and across multiple tissues at the organismal level (rather that individual homogeneous cell lines) is paving the way for a promising decade ahead in translating basic discoveries into interventions that promote lifelong health.

Acknowledgments

We thank Brian Gilmour, Linda Bergersen, and Evandro Fei Fang for inviting us to write this chapter, and apologize to all of the studies and groups we did not reference owing to space constraints. The Samant Lab is funded by Institute Strategic Program Grant BB/P013384/1 from the UK Biotechnology and Biological Sciences Research Council (BBSRC).

References

1. Balch WE, Morimoto RI, Dillin A, Kelly JW. Adapting proteostasis for disease intervention. *Science*. February 15, 2008;319(5865):916—919. https://doi.org/10.1126/science.1141448.
2. Lopez-Otin C, Blasco MA, Partridge L, Serrano M, Kroemer G. The hallmarks of aging. *Cell*. June 6, 2013;153(6):1194—1217. https://doi.org/10.1016/j.cell.2013.05.039.
3. Hutt DM, Powers ET, Balch WE. The proteostasis boundary in misfolding diseases of membrane traffic. *FEBS Lett*. August 20, 2009;583(16):2639—2646. https://doi.org/10.1016/j.febslet.2009.07.014.
4. Powers ET, Morimoto RI, Dillin A, Kelly JW, Balch WE. Biological and chemical approaches to diseases of proteostasis deficiency. *Annu Rev Biochem*. 2009;78:959—991. https://doi.org/10.1146/annurev.biochem.052308.114844.
5. Hartl FU, Bracher A, Hayer-Hartl M. Molecular chaperones in protein folding and proteostasis. *Nature*. July 20, 2011;475(7356):324—332. https://doi.org/10.1038/nature10317.
6. Roth DM, Balch WE. Modeling general proteostasis: proteome balance in health and disease. *Curr Opin Cell Biol*. April 2011;23(2):126—134. https://doi.org/10.1016/j.ceb.2010.11.001.
7. Taylor RC, Dillin A. Aging as an event of proteostasis collapse. *Cold Spring Harbor Perspect Biol*. May 1, 2011;3(5). https://doi.org/10.1101/cshperspect.a004440.
8. Kim YE, Hipp MS, Bracher A, Hayer-Hartl M, Hartl FU. Molecular chaperone functions in protein folding and proteostasis. *Annu Rev Biochem*. 2013;82:323—355. https://doi.org/10.1146/annurev-biochem-060208-092442.
9. Gidalevitz T, Prahlad V, Morimoto RI. The stress of protein misfolding: from single cells to multicellular organisms. *Cold Spring Harbor Perspect Biol*. June 1, 2011;3(6). https://doi.org/10.1101/cshperspect.a009704.
10. Aman Y, Schmauck-Medina T, Hansen M, et al. Autophagy in healthy aging and disease. *Nat Aging*. August 2021;1(8):634—650. https://doi.org/10.1038/s43587-021-00098-4.
11. Gems D, de Magalhaes JP. The hoverfly and the wasp: a critique of the hallmarks of aging as a paradigm. *Ageing Res Rev*. September 2021;70:101407. https://doi.org/10.1016/j.arr.2021.101407.
12. Lopez-Otin C, Kroemer G. Hallmarks of health. *Cell*. January 7, 2021;184(1):33—63. https://doi.org/10.1016/j.cell.2020.11.034.
13. Mittelbrunn M, Kroemer G. Hallmarks of T cell aging. *Nat Immunol*. June 2021;22(6):687—698. https://doi.org/10.1038/s41590-021-00927-z.
14. Kennedy BK, Berger SL, Brunet A, et al. Geroscience: linking aging to chronic disease. *Cell*. November 6, 2014;159(4):709—713. https://doi.org/10.1016/j.cell.2014.10.039.
15. Uhlen M, Fagerberg L, Hallstrom BM, et al. Proteomics. Tissue-based map of the human proteome. *Science*. January 23, 2015;347(6220):1260419. https://doi.org/10.1126/science.1260419.
16. Harper JW, Bennett EJ. Proteome complexity and the forces that drive proteome imbalance. *Nature*. September 15, 2016;537(7620):328—338. https://doi.org/10.1038/nature19947.

17. Milo R. What is the total number of protein molecules per cell volume? A call to rethink some published values. *Bioessays.* December 2013;35(12):1050—1055. https://doi.org/10.1002/bies.201300066.
18. Geiger T, Wehner A, Schaab C, Cox J, Mann M. Comparative proteomic analysis of eleven common cell lines reveals ubiquitous but varying expression of most proteins. *Mol Cell Proteom.* March 2012;11(3). https://doi.org/10.1074/mcp.M111.014050. M111 014050.
19. Chiti F, Dobson CM. Protein misfolding, amyloid formation, and human disease: a summary of progress over the last decade. *Annu Rev Biochem.* June 20, 2017;86:27—68. https://doi.org/10.1146/annurev-biochem-061516-045115.
20. Valastyan JS, Lindquist S. Mechanisms of protein-folding diseases at a glance. *Dis Model Mech.* January 2014;7(1):9—14. https://doi.org/10.1242/dmm.013474.
21. Klaips CL, Jayaraj GG, Hartl FU. Pathways of cellular proteostasis in aging and disease. *J Cell Biol.* January 2, 2018;217(1):51—63. https://doi.org/10.1083/jcb.201709072.
22. Ingolia NT, Lareau LF, Weissman JS. Ribosome profiling of mouse embryonic stem cells reveals the complexity and dynamics of mammalian proteomes. *Cell.* November 11, 2011;147(4):789—802. https://doi.org/10.1016/j.cell.2011.10.002.
23. Joazeiro CAP. Mechanisms and functions of ribosome-associated protein quality control. *Nat Rev Mol Cell Biol.* June 2019;20(6):368—383. https://doi.org/10.1038/s41580-019-0118-2.
24. Inada T. Quality controls induced by aberrant translation. *Nucleic Acids Res.* February 20, 2020;48(3):1084—1096. https://doi.org/10.1093/nar/gkz1201.
25. Sitron CS, Brandman O. Detection and degradation of stalled nascent chains via ribosome-associated quality control. *Annu Rev Biochem.* June 20, 2020;89:417—442. https://doi.org/10.1146/annurev-biochem-013118-110729.
26. Mohler K, Ibba M. Translational fidelity and mistranslation in the cellular response to stress. *Nat Microbiol.* August 24, 2017;2:17117. https://doi.org/10.1038/nmicrobiol.2017.117.
27. Drummond DA, Wilke CO. The evolutionary consequences of erroneous protein synthesis. *Nat Rev Genet.* October 2009;10(10):715—724. https://doi.org/10.1038/nrg2662.
28. Zaher HS, Green R. Fidelity at the molecular level: lessons from protein synthesis. *Cell.* February 20, 2009;136(4):746—762. https://doi.org/10.1016/j.cell.2009.01.036.
29. Vabulas RM, Hartl FU. Protein synthesis upon acute nutrient restriction relies on proteasome function. *Science.* December 23, 2005;310(5756):1960—1963. https://doi.org/10.1126/science.1121925.
30. Schubert U, Anton LC, Gibbs J, Norbury CC, Yewdell JW, Bennink JR. Rapid degradation of a large fraction of newly synthesized proteins by proteasomes. *Nature.* April 13, 2000;404(6779):770—774. https://doi.org/10.1038/35008096.
31. Duttler S, Pechmann S, Frydman J. Principles of cotranslational ubiquitination and quality control at the ribosome. *Mol Cell.* May 9, 2013;50(3):379—393. https://doi.org/10.1016/j.molcel.2013.03.010.
32. Wang F, Durfee LA, Huibregtse JM. A cotranslational ubiquitination pathway for quality control of misfolded proteins. *Mol Cell.* May 9, 2013;50(3):368—378. https://doi.org/10.1016/j.molcel.2013.03.009.
33. Wolff S, Weissman JS, Dillin A. Differential scales of protein quality control. *Cell.* March 27, 2014;157(1):52—64. https://doi.org/10.1016/j.cell.2014.03.007.
34. Wright PE, Dyson HJ. Intrinsically disordered proteins in cellular signalling and regulation. *Nat Rev Mol Cell Biol.* January 2015;16(1):18—29. https://doi.org/10.1038/nrm3920.
35. Anfinsen CB. Principles that govern the folding of protein chains. *Science.* July 20, 1973;181(4096):223—230. https://doi.org/10.1126/science.181.4096.223.

36. Ellis RJ. Macromolecular crowding: an important but neglected aspect of the intracellular environment. *Curr Opin Struct Biol*. February 2001;11(1):114−119. https://doi.org/10.1016/s0959-440x(00)00172-x.

37. Ellis RJ, Hartl FU. Principles of protein folding in the cellular environment. *Curr Opin Struct Biol*. February 1999;9(1):102−110. https://doi.org/10.1016/s0959-440x(99)80013-x.

38. Frydman J. Folding of newly translated proteins in vivo: the role of molecular chaperones. *Annu Rev Biochem*. 2001;70:603−647. https://doi.org/10.1146/annurev.biochem.70.1.603.

39. Buchner J. Supervising the fold: functional principles of molecular chaperones. *Faseb J*. January 1996;10(1):10−19.

40. Balchin D, Hayer-Hartl M, Hartl FU. Recent advances in understanding catalysis of protein folding by molecular chaperones. *FEBS Lett*. September 2020;594(17):2770−2781. https://doi.org/10.1002/1873-3468.13844.

41. Edkins AL, Boshoff A. General structural and functional features of molecular chaperones. *Adv Exp Med Biol*. 2021;1340:11−73. https://doi.org/10.1007/978-3-030-78397-6_2.

42. Albanese V, Yam AY, Baughman J, Parnot C, Frydman J. Systems analyses reveal two chaperone networks with distinct functions in eukaryotic cells. *Cell*. January 13, 2006;124(1):75−88. https://doi.org/10.1016/j.cell.2005.11.039.

43. Waudby CA, Dobson CM, Christodoulou J. Nature and regulation of protein folding on the ribosome. *Trends Biochem Sci*. November 2019;44(11):914−926. https://doi.org/10.1016/j.tibs.2019.06.008.

44. Wilson DN, Beckmann R. The ribosomal tunnel as a functional environment for nascent polypeptide folding and translational stalling. *Curr Opin Struct Biol*. April 2011;21(2):274−282. https://doi.org/10.1016/j.sbi.2011.01.007.

45. Wruck F, Tian P, Kudva R, et al. The ribosome modulates folding inside the ribosomal exit tunnel. *Commun Biol*. May 5, 2021;4(1):523. https://doi.org/10.1038/s42003-021-02055-8.

46. Nilsson OB, Hedman R, Marino J, et al. Cotranslational protein folding inside the ribosome exit tunnel. *Cell Rep*. September 8, 2015;12(10):1533−1540. https://doi.org/10.1016/j.celrep.2015.07.065.

47. Rodnina MV. The ribosome in action: tuning of translational efficiency and protein folding. *Protein Sci*. August 2016;25(8):1390−1406. https://doi.org/10.1002/pro.2950.

48. Wilson DN, Arenz S, Beckmann R. Translation regulation via nascent polypeptide-mediated ribosome stalling. *Curr Opin Struct Biol*. April 2016;37:123−133. https://doi.org/10.1016/j.sbi.2016.01.008.

49. Hartl FU, Hayer-Hartl M. Molecular chaperones in the cytosol: from nascent chain to folded protein. *Science*. March 8, 2002;295(5561):1852−1858. https://doi.org/10.1126/science.1068408.

50. Deuerling E, Bukau B. Chaperone-assisted folding of newly synthesized proteins in the cytosol. *Crit Rev Biochem Mol Biol*. Sep-Dec 2004;39(5−6):261−277. https://doi.org/10.1080/10409230490892496.

51. Kohler V, Andreasson C. Hsp70-mediated quality control: should I stay or should I go? *Biol Chem*. October 25, 2020;401(11):1233−1248. https://doi.org/10.1515/hsz-2020-0187.

52. Mayer MP, Gierasch LM. Recent advances in the structural and mechanistic aspects of Hsp70 molecular chaperones. *J Biol Chem*. February 8, 2019;294(6):2085−2097. https://doi.org/10.1074/jbc.REV118.002810.

53. Rosenzweig R, Nillegoda NB, Mayer MP, Bukau B. The Hsp70 chaperone network. *Nat Rev Mol Cell Biol*. November 2019;20(11):665−680. https://doi.org/10.1038/s41580-019-0133-3.

54. Gestaut D, Limatola A, Joachimiak L, Frydman J. The ATP-powered gymnastics of TRiC/CCT: an asymmetric protein folding machine with a symmetric origin story. *Curr Opin Struct Biol.* April 2019;55:50—58. https://doi.org/10.1016/j.sbi.2019.03.002.

55. Schopf FH, Biebl MM, Buchner J. The HSP90 chaperone machinery. *Nat Rev Mol Cell Biol.* June 2017;18(6):345—360. https://doi.org/10.1038/nrm.2017.20.

56. Baldwin AJ, Knowles TP, Tartaglia GG, et al. Metastability of native proteins and the phenomenon of amyloid formation. *J Am Chem Soc.* September 14, 2011;133(36):14160—14163. https://doi.org/10.1021/ja2017703.

57. Ghosh DK, Ranjan A. The metastable states of proteins. *Protein Sci.* July 2020;29(7):1559—1568. https://doi.org/10.1002/pro.3859.

58. Gershenson A, Gierasch LM, Pastore A, Radford SE. Energy landscapes of functional proteins are inherently risky. *Nat Chem Biol.* November 2014;10(11):884—891. https://doi.org/10.1038/nchembio.1670.

59. Tissieres A, Mitchell HK, Tracy UM. Protein synthesis in salivary glands of *Drosophila melanogaster*: relation to chromosome puffs. *J Mol Biol.* April 15, 1974;84(3):389—398. https://doi.org/10.1016/0022-2836(74)90447-1.

60. Dean ME, Johnson JL. Human Hsp90 cochaperones: perspectives on tissue-specific expression and identification of cochaperones with similar in vivo functions. *Cell Stress Chaperones.* January 2021;26(1):3—13. https://doi.org/10.1007/s12192-020-01167-0.

61. Kampinga HH, Craig EA. The HSP70 chaperone machinery: J proteins as drivers of functional specificity. *Nat Rev Mol Cell Biol.* August 2010;11(8):579—592. https://doi.org/10.1038/nrm2941.

62. Webster JM, Darling AL, Uversky VN, Blair LJ. Small heat shock proteins, big impact on protein aggregation in neurodegenerative disease. *Front Pharmacol.* 2019;10:1047. https://doi.org/10.3389/fphar.2019.01047.

63. Haslbeck M, Weinkauf S, Buchner J. Small heat shock proteins: simplicity meets complexity. *J Biol Chem.* February 8, 2019;294(6):2121—2132. https://doi.org/10.1074/jbc.REV118.002809.

64. Sontag EM, Samant RS, Frydman J. Mechanisms and functions of spatial protein quality control. *Annu Rev Biochem.* June 20, 2017;86:97—122. https://doi.org/10.1146/annurev-biochem-060815-014616.

65. Miller SB, Mogk A, Bukau B. Spatially organized aggregation of misfolded proteins as cellular stress defense strategy. *J Mol Biol.* April 10, 2015;427(7):1564—1574. https://doi.org/10.1016/j.jmb.2015.02.006.

66. Escusa-Toret S, Vonk WI, Frydman J. Spatial sequestration of misfolded proteins by a dynamic chaperone pathway enhances cellular fitness during stress. *Nat Cell Biol.* October 2013;15(10):1231—1243. https://doi.org/10.1038/ncb2838.

67. Johnston HE, Samant RS. Alternative systems for misfolded protein clearance: life beyond the proteasome. *FEBS J.* November 1, 2020. https://doi.org/10.1111/febs.15617.

68. Esser C, Alberti S, Hohfeld J. Cooperation of molecular chaperones with the ubiquitin/proteasome system. *Biochim Biophys Acta.* November 29, 2004;1695(1—3):171—188. https://doi.org/10.1016/j.bbamcr.2004.09.020.

69. Itakura E, Zavodszky E, Shao S, Wohlever ML, Keenan RJ, Hegde RS. Ubiquilins chaperone and triage mitochondrial membrane proteins for degradation. *Mol Cell.* July 7, 2016;63(1):21—33. https://doi.org/10.1016/j.molcel.2016.05.020.

70. Shao S, Rodrigo-Brenni MC, Kivlen MH, Hegde RS. Mechanistic basis for a molecular triage reaction. *Science.* January 20, 2017;355(6322):298—302. https://doi.org/10.1126/science.aah6130.

71. Sturner E, Behl C. The role of the multifunctional BAG3 protein in cellular protein quality control and in disease. *Front Mol Neurosci.* 2017;10:177. https://doi.org/10.3389/fnmol.2017.00177.

72. Lu K, den Brave F, Jentsch S. Receptor oligomerization guides pathway choice between proteasomal and autophagic degradation. *Nat Cell Biol.* June 2017;19(6):732–739. https://doi.org/10.1038/ncb3531.

73. Bard JAM, Goodall EA, Greene ER, Jonsson E, Dong KC, Martin A. Structure and function of the 26S proteasome. *Annu Rev Biochem.* June 20, 2018;87:697–724. https://doi.org/10.1146/annurev-biochem-062917-011931.

74. Marshall RS, Vierstra RD. Dynamic regulation of the 26S proteasome: from synthesis to degradation. Review. *Front Mol Biosci.* 2019-June-07;6(40). https://doi.org/10.3389/fmolb.2019.00040, 2019.

75. Akutsu M, Dikic I, Bremm A. Ubiquitin chain diversity at a glance. *J Cell Sci.* March 1, 2016;129(5):875–880. https://doi.org/10.1242/jcs.183954.

76. Kwon YT, Ciechanover A. The ubiquitin code in the ubiquitin-proteasome system and autophagy. *Trends Biochem Sci.* November 2017;42(11):873–886. https://doi.org/10.1016/j.tibs.2017.09.002.

77. Komander D, Rape M. The ubiquitin code. *Annu Rev Biochem.* 2012;81:203–229. https://doi.org/10.1146/annurev-biochem-060310-170328.

78. Samant RS, Livingston CM, Sontag EM, Frydman J. Distinct proteostasis circuits cooperate in nuclear and cytoplasmic protein quality control. *Nature.* November 2018;563(7731):407–411. https://doi.org/10.1038/s41586-018-0678-x.

79. Yau RG, Doerner K, Castellanos ER, et al. Assembly and function of heterotypic ubiquitin chains in cell-cycle and protein quality control. *Cell.* November 2, 2017;171(4):918–933 e20. https://doi.org/10.1016/j.cell.2017.09.040.

80. Leto DE, Morgens DW, Zhang L, et al. Genome-wide CRISPR analysis identifies substrate-specific conjugation modules in ER-associated degradation. *Mol Cell.* January 17, 2019;73(2):377–389 e11. https://doi.org/10.1016/j.molcel.2018.11.015.

81. Hershko A, Ciechanover A. The ubiquitin system. *Annu Rev Biochem.* 1998;67:425–479. https://doi.org/10.1146/annurev.biochem.67.1.425.

82. Cappadocia L, Lima CD. Ubiquitin-like protein conjugation: structures, chemistry, and mechanism. *Chem Rev.* February 14, 2018;118(3):889–918. https://doi.org/10.1021/acs.chemrev.6b00737.

83. Joshi V, Amanullah A, Upadhyay A, Mishra R, Kumar A, Mishra A. A decade of boon or burden: what has the CHIP ever done for cellular protein quality control mechanism implicated in neurodegeneration and aging? *Front Mol Neurosci.* 2016;9:93. https://doi.org/10.3389/fnmol.2016.00093.

84. Needham PG, Guerriero CJ, Brodsky JL. Chaperoning endoplasmic reticulum-associated degradation (ERAD) and protein conformational diseases. *Cold Spring Harbor Perspect Biol.* August 1, 2019;11(8). https://doi.org/10.1101/cshperspect.a033928.

85. Wu X, Rapoport TA. Mechanistic insights into ER-associated protein degradation. *Curr Opin Cell Biol.* August 2018;53:22–28. https://doi.org/10.1016/j.ceb.2018.04.004.

86. van den Boom J, Meyer H. VCP/p97-Mediated unfolding as a principle in protein homeostasis and signaling. *Mol Cell.* January 18, 2018;69(2):182–194. https://doi.org/10.1016/j.molcel.2017.10.028.

87. Fleming A, Bourdenx M, Fujimaki M, et al. The different autophagy degradation pathways and neurodegeneration. *Neuron.* 2022. https://doi.org/10.1016/j.neuron.2022.01.017, 2022/02/07/.

88. Lawrence RE, Zoncu R. The lysosome as a cellular centre for signalling, metabolism and quality control. *Nat Cell Biol.* February 2019;21(2):133–142. https://doi.org/10.1038/s41556-018-0244-7.

89. Johansen T, Lamark T. Selective autophagy: ATG8 family proteins, LIR motifs and cargo receptors. *J Mol Biol.* January 3, 2020;432(1):80–103. https://doi.org/10.1016/j.jmb.2019.07.016.

90. Grumati P, Dikic I. Ubiquitin signaling and autophagy. *J Biol Chem.* April 13, 2018;293(15):5404–5413. https://doi.org/10.1074/jbc.TM117.000117.

91. Zellner S, Schifferer M, Behrends C. Systematically defining selective autophagy receptor-specific cargo using autophagosome content profiling. *Mol Cell*. March 18, 2021;81(6):1337−1354 e8. https://doi.org/10.1016/j.molcel.2021.01.009.

92. Lu K, Psakhye I, Jentsch S. Autophagic clearance of polyQ proteins mediated by ubiquitin-Atg8 adaptors of the conserved CUET protein family. *Cell*. July 31, 2014;158(3):549−563. https://doi.org/10.1016/j.cell.2014.05.048.

93. Biel TG, Aryal B, Gerber MH, Trevino JG, Mizuno N, Rao VA. Mitochondrial dysfunction generates aggregates that resist lysosomal degradation in human breast cancer cells. *Cell Death Dis*. June 15, 2020;11(6):460. https://doi.org/10.1038/s41419-020-2658-y.

94. Korac J, Schaeffer V, Kovacevic I, et al. Ubiquitin-independent function of optineurin in autophagic clearance of protein aggregates. *J Cell Sci*. January 15, 2013;126(Pt 2):580−592. https://doi.org/10.1242/jcs.114926.

95. Sarraf SA, Shah HV, Kanfer G, et al. Loss of TAX1BP1-directed autophagy results in protein aggregate accumulation in the brain. *Mol Cell*. December 3, 2020;80(5):779−795 e10. https://doi.org/10.1016/j.molcel.2020.10.041.

96. Heckmann BL, Green DR. LC3-associated phagocytosis at a glance. *J Cell Sci*. February 20, 2019;132(5). https://doi.org/10.1242/jcs.222984.

97. Heckmann BL, Teubner BJW, Tummers B, et al. LC3-Associated endocytosis facilitates beta-amyloid clearance and mitigates neurodegeneration in murine alzheimer's disease. *Cell*. July 25, 2019;178(3):536−551 e14. https://doi.org/10.1016/j.cell.2019.05.056.

98. Itakura E, Chiba M, Murata T, Matsuura A. Heparan sulfate is a clearance receptor for aberrant extracellular proteins. *J Cell Biol*. March 2, 2020;219(3). https://doi.org/10.1083/jcb.201911126.

99. Kaganovich D, Kopito R, Frydman J. Misfolded proteins partition between two distinct quality control compartments. *Nature*. August 28, 2008;454(7208):1088−1095. https://doi.org/10.1038/nature07195.

100. Labbadia J, Morimoto RI. The biology of proteostasis in aging and disease. *Annu Rev Biochem*. 2015;84:435−464. https://doi.org/10.1146/annurev-biochem-060614-033955.

101. Hipp MS, Kasturi P, Hartl FU. The proteostasis network and its decline in ageing. *Nat Rev Mol Cell Biol*. July 2019;20(7):421−435. https://doi.org/10.1038/s41580-019-0101-y.

102. Thibaudeau TA, Anderson RT, Smith DM. A common mechanism of proteasome impairment by neurodegenerative disease-associated oligomers. *Nat Commun*. March 15, 2018;9(1):1097. https://doi.org/10.1038/s41467-018-03509-0.

103. Moore DL, Jessberger S. Creating age asymmetry: consequences of inheriting damaged goods in mammalian cells. *Trends Cell Biol*. January 2017;27(1):82−92. https://doi.org/10.1016/j.tcb.2016.09.007.

104. Kysela DT, Brown PJ, Huang KC, Brun YV. Biological consequences and advantages of asymmetric bacterial growth. *Annu Rev Microbiol*. 2013;67:417−435. https://doi.org/10.1146/annurev-micro-092412-155622.

105. Hill SM, Hanzen S, Nystrom T. Restricted access: spatial sequestration of damaged proteins during stress and aging. *EMBO Rep*. March 2017;18(3):377−391. https://doi.org/10.15252/embr.201643458.

106. Loeffler D, Wehling A, Schneiter F, et al. Asymmetric lysosome inheritance predicts activation of haematopoietic stem cells. *Nature*. September 2019;573(7774):426−429. https://doi.org/10.1038/s41586-019-1531-6.

107. Ogrodnik M, Salmonowicz H, Brown R, et al. Dynamic JUNQ inclusion bodies are asymmetrically inherited in mammalian cell lines through the asymmetric partitioning of vimentin. *Proc Natl Acad Sci U S A*. June 3, 2014;111(22):8049−8054. https://doi.org/10.1073/pnas.1324035111.

108. Vedel S, Nunns H, Kosmrlj A, Semsey S, Trusina A. Asymmetric damage segregation constitutes an emergent population-level stress response. *Cell Syst.* August 2016;3(2):187−198. https://doi.org/10.1016/j.cels.2016.06.008.

109. Lindner AB, Madden R, Demarez A, Stewart EJ, Taddei F. Asymmetric segregation of protein aggregates is associated with cellular aging and rejuvenation. *Proc Natl Acad Sci U S A.* February 26, 2008;105(8):3076−3081. https://doi.org/10.1073/pnas.0708931105.

110. Zhang T, Wolfe C, Pierle A, Welle KA, Hryhorenko JR, Ghaemmaghami S. Proteome-wide modulation of degradation dynamics in response to growth arrest. *Proc Natl Acad Sci U S A.* November 28, 2017;114(48):E10329−E10338. https://doi.org/10.1073/pnas.1710238114.

111. Lee HJ, Cho ED, Lee KW, Kim JH, Cho SG, Lee SJ. Autophagic failure promotes the exocytosis and intercellular transfer of alpha-synuclein. *Exp Mol Med.* May 10, 2013;45:e22. https://doi.org/10.1038/emm.2013.45.

112. Lee JG, Takahama S, Zhang G, Tomarev SI, Ye Y. Unconventional secretion of misfolded proteins promotes adaptation to proteasome dysfunction in mammalian cells. *Nat Cell Biol.* July 2016;18(7):765−776. https://doi.org/10.1038/ncb3372.

113. Fontaine SN, Zheng D, Sabbagh JJ, et al. DnaJ/Hsc70 chaperone complexes control the extracellular release of neurodegenerative-associated proteins. *EMBO J.* July 15, 2016;35(14):1537−1549. https://doi.org/10.15252/embj.201593489.

114. Deng J, Koutras C, Donnelier J, et al. Neurons export extracellular vesicles enriched in cysteine string protein and misfolded protein cargo. *Sci Rep.* April 19, 2017;7(1):956. https://doi.org/10.1038/s41598-017-01115-6.

115. Xu Y, Cui L, Dibello A, et al. DNAJC5 facilitates USP19-dependent unconventional secretion of misfolded cytosolic proteins. *Cell Discov.* 2018;4:11. https://doi.org/10.1038/s41421-018-0012-7.

116. Ciechanover A, Kwon YT. Protein quality control by molecular chaperones in neurodegeneration. *Front Neurosci.* 2017;11:185. https://doi.org/10.3389/fnins.2017.00185.

117. Henning RH, Brundel B. Proteostasis in cardiac health and disease. *Nat Rev Cardiol.* November 2017;14(11):637−653. https://doi.org/10.1038/nrcardio.2017.89.

118. Kong KE, Coelho JPL, Feige MJ, Khmelinskii A. Quality control of mislocalized and orphan proteins. *Exp Cell Res.* June 15, 2021;403(2):112617. https://doi.org/10.1016/j.yexcr.2021.112617.

119. Sommer MS, Schleiff E. Protein targeting and transport as a necessary consequence of increased cellular complexity. *Cold Spring Harbor Perspect Biol.* August 1, 2014;6(8). https://doi.org/10.1101/cshperspect.a016055.

120. Natan E, Wells JN, Teichmann SA, Marsh JA. Regulation, evolution and consequences of cotranslational protein complex assembly. *Curr Opin Struct Biol.* February 2017;42:90−97. https://doi.org/10.1016/j.sbi.2016.11.023.

121. Thul PJ, Akesson L, Wiking M, et al. A subcellular map of the human proteome. *Science.* May 26, 2017;356(6340). https://doi.org/10.1126/science.aal3321.

122. Geladaki A, Kocevar Britovsek N, Breckels LM, et al. Combining LOPIT with differential ultracentrifugation for high-resolution spatial proteomics. *Nat Commun.* January 18, 2019;10(1):331. https://doi.org/10.1038/s41467-018-08191-w.

123. Kim DH, Hwang I. Direct targeting of proteins from the cytosol to organelles: the ER versus endosymbiotic organelles. *Traffic.* June 2013;14(6):613−621. https://doi.org/10.1111/tra.12043.

124. Hegde RS, Zavodszky E. Recognition and degradation of mislocalized proteins in health and disease. *Cold Spring Harbor Perspect Biol.* November 1, 2019;11(11). https://doi.org/10.1101/cshperspect.a033902.

125. Hutt DM, Balch WE. Expanding proteostasis by membrane trafficking networks. *Cold Spring Harbor Perspect Biol.* July 1, 2013;5(7). https://doi.org/10.1101/cshperspect.a013383.

126. Broix L, Turchetto S, Nguyen L. Coordination between transport and local translation in neurons. *Trends Cell Biol.* May 2021;31(5):372–386. https://doi.org/10.1016/j.tcb.2021.01.001.

127. Genuth NR, Barna M. The discovery of ribosome heterogeneity and its implications for gene regulation and organismal life. *Mol Cell.* August 2, 2018;71(3):364–374. https://doi.org/10.1016/j.molcel.2018.07.018.

128. Schwarz A, Beck M. The benefits of cotranslational assembly: a structural perspective. *Trends Cell Biol.* October 2019;29(10):791–803. https://doi.org/10.1016/j.tcb.2019.07.006.

129. Shiber A, Doring K, Friedrich U, et al. Cotranslational assembly of protein complexes in eukaryotes revealed by ribosome profiling. *Nature.* September 2018;561(7722):268–272. https://doi.org/10.1038/s41586-018-0462-y.

130. Ishikawa K. Multilayered regulation of proteome stoichiometry. *Curr Genet.* December 2021;67(6):883–890. https://doi.org/10.1007/s00294-021-01205-z.

131. Juszkiewicz S, Hegde RS. Quality control of orphaned proteins. *Mol Cell.* August 2, 2018;71(3):443–457. https://doi.org/10.1016/j.molcel.2018.07.001.

132. Natarajan N, Foresti O, Wendrich K, Stein A, Carvalho P. Quality control of protein complex assembly by a transmembrane recognition factor. *Mol Cell.* January 2, 2020;77(1):108–119 e9. https://doi.org/10.1016/j.molcel.2019.10.003.

133. Yanagitani K, Juszkiewicz S, Hegde RS. UBE2O is a quality control factor for orphans of multiprotein complexes. *Science.* August 4, 2017;357(6350):472–475. https://doi.org/10.1126/science.aan0178.

134. Miyazawa M, Tashiro E, Kitaura H, et al. Prefoldin subunits are protected from ubiquitin-proteasome system-mediated degradation by forming complex with other constituent subunits. *J Biol Chem.* June 3, 2011;286(22):19191–19203. https://doi.org/10.1074/jbc.M110.216259.

135. Zavodszky E, Peak-Chew SY, Juszkiewicz S, Narvaez AJ, Hegde RS. Identification of a quality-control factor that monitors failures during proteasome assembly. *Science.* August 27, 2021;373(6558):998–1004. https://doi.org/10.1126/science.abc6500.

136. Taggart JC, Zauber H, Selbach M, Li GW, McShane E. Keeping the proportions of protein complex components in check. *Cell Syst.* February 26, 2020;10(2):125–132. https://doi.org/10.1016/j.cels.2020.01.004.

137. Liu Y, Borel C, Li L, et al. Systematic proteome and proteostasis profiling in human Trisomy 21 fibroblast cells. *Nat Commun.* October 31, 2017;8(1):1212. https://doi.org/10.1038/s41467-017-01422-6.

138. Brennan CM, Vaites LP, Wells JN, et al. Protein aggregation mediates stoichiometry of protein complexes in aneuploid cells. *Genes Dev.* August 1, 2019;33(15–16):1031–1047. https://doi.org/10.1101/gad.327494.119.

139. Joy J, Barrio L, Santos-Tapia C, et al. Proteostasis failure and mitochondrial dysfunction leads to aneuploidy-induced senescence. *Dev Cell.* July 26, 2021;56(14):2043–2058 e7. https://doi.org/10.1016/j.devcel.2021.06.009.

140. Samant RS, Masto VB, Frydman J. Dosage compensation plans: protein aggregation provides additional insurance against aneuploidy. *Genes Dev.* August 1, 2019;33(15–16):1027–1030. https://doi.org/10.1101/gad.329383.119.

141. Joutsen J, Sistonen L. Tailoring of proteostasis networks with heat shock factors. *Cold Spring Harbor Perspect Biol.* April 1, 2019;11(4). https://doi.org/10.1101/cshperspect.a034066.

142. Costa-Mattioli M, Walter P. The integrated stress response: from mechanism to disease. *Science.* April 24, 2020;368(6489). https://doi.org/10.1126/science.aat5314.

143. Hetz C, Zhang K, Kaufman RJ. Mechanisms, regulation and functions of the unfolded protein response. *Nat Rev Mol Cell Biol.* August 2020;21(8):421–438. https://doi.org/10.1038/s41580-020-0250-z.

144. Shpilka T, Haynes CM. The mitochondrial UPR: mechanisms, physiological functions and implications in ageing. *Nat Rev Mol Cell Biol.* February 2018;19(2):109—120. https://doi.org/10.1038/nrm.2017.110.
145. Bellezza I, Giambanco I, Minelli A, Donato R. Nrf2-Keap1 signaling in oxidative and reductive stress. *Biochim Biophys Acta Mol Cell Res.* May 2018;1865(5):721—733. https://doi.org/10.1016/j.bbamcr.2018.02.010.
146. Ma Q. Role of nrf2 in oxidative stress and toxicity. *Annu Rev Pharmacol Toxicol.* 2013;53:401—426. https://doi.org/10.1146/annurev-pharmtox-011112-140320.
147. Santiago AM, Goncalves DL, Morano KA. Mechanisms of sensing and response to proteotoxic stress. *Exp Cell Res.* October 15, 2020;395(2):112240. https://doi.org/10.1016/j.yexcr.2020.112240.
148. Masser AE, Ciccarelli M, Andreasson C. Hsf1 on a leash - controlling the heat shock response by chaperone titration. *Exp Cell Res.* November 1, 2020;396(1):112246. https://doi.org/10.1016/j.yexcr.2020.112246.
149. Amin-Wetzel N, Neidhardt L, Yan Y, Mayer MP, Ron D. Unstructured regions in IRE1alpha specify BiP-mediated destabilisation of the luminal domain dimer and repression of the UPR. *Elife.* December 24, 2019;8. https://doi.org/10.7554/eLife.50793.
150. Karagoz GE, Acosta-Alvear D, Walter P. The unfolded protein response: detecting and responding to fluctuations in the protein-folding capacity of the endoplasmic reticulum. *Cold Spring Harbor Perspect Biol.* September 3, 2019;11(9). https://doi.org/10.1101/cshperspect.a033886.
151. Liu B, Han Y, Qian SB. Cotranslational response to proteotoxic stress by elongation pausing of ribosomes. *Mol Cell.* February 7, 2013;49(3):453—463. https://doi.org/10.1016/j.molcel.2012.12.001.
152. Jucker M, Walker LC. Propagation and spread of pathogenic protein assemblies in neurodegenerative diseases. *Nat Neurosci.* October 2018;21(10):1341—1349. https://doi.org/10.1038/s41593-018-0238-6.
153. Miles J, Scherz-Shouval R, van Oosten-Hawle P. Expanding the organismal proteostasis network: linking systemic stress signaling with the innate immune response. *Trends Biochem Sci.* November 2019;44(11):927—942. https://doi.org/10.1016/j.tibs.2019.06.009.
154. Sala AJ, Bott LC, Morimoto RI. Shaping proteostasis at the cellular, tissue, and organismal level. *J Cell Biol.* May 1, 2017;216(5):1231—1241. https://doi.org/10.1083/jcb.201612111.
155. Prahlad V. The discovery and consequences of the central role of the nervous system in the control of protein homeostasis. *J Neurogenet.* Sep-Dec 2020;34(3—4):489—499. https://doi.org/10.1080/01677063.2020.1771333.
156. Gallotta I, Sandhu A, Peters M, et al. Extracellular proteostasis prevents aggregation during pathogenic attack. *Nature.* August 2020;584(7821):410—414. https://doi.org/10.1038/s41586-020-2461-z.
157. Tian X, Seluanov A, Gorbunova V. Molecular mechanisms determining lifespan in short- and long-lived species. *Trends Endocrinol Metabol.* October 2017;28(10):722—734. https://doi.org/10.1016/j.tem.2017.07.004.
158. Ke Z, Mallik P, Johnson AB, et al. Translation fidelity coevolves with longevity. *Aging Cell.* October 2017;16(5):988—993. https://doi.org/10.1111/acel.12628.
159. Draceni Y, Pechmann S. Pervasive convergent evolution and extreme phenotypes define chaperone requirements of protein homeostasis. *Proc Natl Acad Sci U S A.* October 1, 2019;116(40):20009—20014. https://doi.org/10.1073/pnas.1904611116.
160. Powers ET, Balch WE. Diversity in the origins of proteostasis networks—a driver for protein function in evolution. *Nat Rev Mol Cell Biol.* April 2013;14(4):237—248. https://doi.org/10.1038/nrm3542.

161. Pride H, Yu Z, Sunchu B, et al. Long-lived species have improved proteostasis compared to phylogenetically-related shorter-lived species. *Biochem Biophys Res Commun.* February 20, 2015;457(4):669–675. https://doi.org/10.1016/j.bbrc.2015.01.046.

162. Swovick K, Firsanov D, Welle KA, et al. Interspecies differences in proteome turnover kinetics are correlated with life spans and energetic demands. *Mol Cell Proteomics.* 2021;20:100041. https://doi.org/10.1074/mcp.RA120.002301.

163. Lewis KN, Mele J, Hornsby PJ, Buffenstein R. Stress resistance in the naked mole-rat: the bare essentials - a mini-review. *Gerontology.* 2012;58(5):453–462. https://doi.org/10.1159/000335966.

164. Ruby JG, Smith M, Buffenstein R. Naked Mole-Rat mortality rates defy gompertzian laws by not increasing with age. *Elife.* January 24, 2018;7. https://doi.org/10.7554/eLife.31157.

165. Ruano D. Proteostasis dysfunction in aged mammalian cells. The stressful role of inflammation. *Front Mol Biosci.* 2021;8:658742. https://doi.org/10.3389/fmolb.2021.658742.

166. Komljenovic A, Li H, Sorrentino V, Kutalik Z, Auwerx J, Robinson-Rechavi M. Cross-species functional modules link proteostasis to human normal aging. *PLoS Comput Biol.* July 2019;15(7):e1007162. https://doi.org/10.1371/journal.pcbi.1007162.

167. Javidnia S, Cranwell S, Mueller SH, et al. Mendelian randomization analyses implicate biogenesis of translation machinery in human aging. *Genome Res.* February 2022;32(2):258–265. https://doi.org/10.1101/gr.275636.121.

168. Hadizadeh Esfahani A, Sverchkova A, Saez-Rodriguez J, Schuppert AA, Brehme M. A systematic atlas of chaperome deregulation topologies across the human cancer landscape. *PLoS Comput Biol.* January 2018;14(1):e1005890. https://doi.org/10.1371/journal.pcbi.1005890.

169. Jaeger AM, Whitesell L. HSP90: enabler of cancer adaptation. *Annu Rev Cell Biol.* 2019;3(1):275–297. https://doi.org/10.1146/annurev-cancerbio-030518-055533.

170. Ramsey JJ, Harper ME, Weindruch R. Restriction of energy intake, energy expenditure, and aging. *Free Radic Biol Med.* November 15, 2000;29(10):946–968. https://doi.org/10.1016/s0891-5849(00)00417-2.

171. Gladyshev VN. The free radical theory of aging is dead. Long live the damage theory. *Antioxidants Redox Signal.* February 1, 2014;20(4):727–731. https://doi.org/10.1089/ars.2013.5228.

172. Maklakov AA, Chapman T. Evolution of ageing as a tangle of trade-offs: energy versus function. *Proc Biol Sci.* September 25, 2019;286(1911):20191604. https://doi.org/10.1098/rspb.2019.1604.

173. Flatt T, Partridge L. Horizons in the evolution of aging. *BMC Biol.* August 20, 2018;16(1):93. https://doi.org/10.1186/s12915-018-0562-z.

174. Labbadia J, Morimoto RI. Repression of the heat shock response is a programmed event at the onset of reproduction. *Mol Cell.* August 20, 2015;59(4):639–650. https://doi.org/10.1016/j.molcel.2015.06.027.

175. Basisty N, Meyer JG, Schilling B. Protein turnover in aging and longevity. *Proteomics.* March 2018;18(5–6):e1700108. https://doi.org/10.1002/pmic.201700108.

176. Anisimova AS, Alexandrov AI, Makarova NE, Gladyshev VN, Dmitriev SE. Protein synthesis and quality control in aging. *Aging.* December 18, 2018;10(12):4269–4288. https://doi.org/10.18632/aging.101721.

177. Gonskikh Y, Polacek N. Alterations of the translation apparatus during aging and stress response. *Mech Ageing Dev.* December 2017;168:30–36. https://doi.org/10.1016/j.mad.2017.04.003.

178. Stein KC, Morales-Polanco F, van der Lienden J, Rainbolt TK, Frydman J. Ageing exacerbates ribosome pausing to disrupt cotranslational proteostasis. *Nature.* 2022/01/19. https://doi.org/10.1038/s41586-021-04295-4, 2022.

179. Kirstein-Miles J, Scior A, Deuerling E, Morimoto RI. The nascent polypeptide-associated complex is a key regulator of proteostasis. *EMBO J.* May 15, 2013;32(10):1451–1468. https://doi.org/10.1038/emboj.2013.87.
180. Kelmer Sacramento E, Kirkpatrick JM, Mazzetto M, et al. Reduced proteasome activity in the aging brain results in ribosome stoichiometry loss and aggregation. *Mol Syst Biol.* June 2020;16(6):e9596. https://doi.org/10.15252/msb.20209596.
181. Santra M, Dill KA, de Graff AMR. Proteostasis collapse is a driver of cell aging and death. *Proc Natl Acad Sci U S A.* October 29, 2019;116(44):22173–22178. https://doi.org/10.1073/pnas.1906592116.
182. Golden TR, Melov S. Microarray analysis of gene expression with age in individual nematodes. *Aging Cell.* June 2004;3(3):111–124. https://doi.org/10.1111/j.1474-9728.2004.00095.x.
183. Walther DM, Kasturi P, Zheng M, et al. Widespread proteome remodeling and aggregation in aging C. elegans. *Cell.* May 7, 2015;161(4):919–932. https://doi.org/10.1016/j.cell.2015.03.032.
184. Anisimova AS, Meerson MB, Gerashchenko MV, Kulakovskiy IV, Dmitriev SE, Gladyshev VN. Multifaceted deregulation of gene expression and protein synthesis with age. *Proc Natl Acad Sci U S A.* July 7, 2020;117(27):15581–15590. https://doi.org/10.1073/pnas.2001788117.
185. Soti C, Csermely P. Molecular chaperones and the aging process. *Biogerontology.* 2000;1(3):225–233. https://doi.org/10.1023/a:1010082129022.
186. Calderwood SK, Murshid A, Prince T. The shock of aging: molecular chaperones and the heat shock response in longevity and aging–a mini-review. *Gerontology.* 2009;55(5):550–558. https://doi.org/10.1159/000225957.
187. Haigis MC, Yankner BA. The aging stress response. *Mol Cell.* October 22, 2010;40(2):333–344. https://doi.org/10.1016/j.molcel.2010.10.002.
188. Gomez CR. Role of heat shock proteins in aging and chronic inflammatory diseases. *Geroscience.* October 2021;43(5):2515–2532. https://doi.org/10.1007/s11357-021-00394-2.
189. Trivedi R, Jurivich DA. A molecular perspective on age-dependent changes to the heat shock axis. *Exp Gerontol.* August 2020;137:110969. https://doi.org/10.1016/j.exger.2020.110969.
190. Brehme M, Voisine C, Rolland T, et al. A chaperome subnetwork safeguards proteostasis in aging and neurodegenerative disease. *Cell Rep.* November 6, 2014;9(3):1135–1150. https://doi.org/10.1016/j.celrep.2014.09.042.
191. Hsu AL, Murphy CT, Kenyon C. Regulation of aging and age-related disease by DAF-16 and heat-shock factor. *Science.* May 16, 2003;300(5622):1142–1145. https://doi.org/10.1126/science.1083701.
192. Kurapati R, Passananti HB, Rose MR, Tower J. Increased hsp22 RNA levels in Drosophila lines genetically selected for increased longevity. *J Gerontol A Biol Sci Med Sci.* November 2000;55(11):B552–B559. https://doi.org/10.1093/gerona/55.11.b552.
193. Vonk WIM, Rainbolt TK, Dolan PT, Webb AE, Brunet A, Frydman J. Differentiation drives widespread rewiring of the neural stem cell chaperone network. *Mol Cell.* April 16, 2020;78(2):329–345 e9. https://doi.org/10.1016/j.molcel.2020.03.009.
194. Rodina A, Wang T, Yan P, et al. The epichaperome is an integrated chaperome network that facilitates tumour survival. *Nature.* October 20, 2016;538(7625):397–401. https://doi.org/10.1038/nature19807.
195. Ginsberg SD, Joshi S, Sharma S, et al. The penalty of stress - epichaperomes negatively reshaping the brain in neurodegenerative disorders. *J Neurochem.* December 2021;159(6):958–979. https://doi.org/10.1111/jnc.15525.
196. Kamal A, Thao L, Sensintaffar J, et al. A high-affinity conformation of Hsp90 confers tumour selectivity on Hsp90 inhibitors. *Nature.* September 25, 2003;425(6956):407–410. https://doi.org/10.1038/nature01913.

197. Brehme M, Sverchkova A, Voisine C. Proteostasis network deregulation signatures as biomarkers for pharmacological disease intervention. *Curr Opin Struct Biol.* 2019;15:74–81. https://doi.org/10.1016/j.coisb.2019.03.008, 2019/06/01/.

198. Saez I, Vilchez D. The mechanistic links between proteasome activity, aging and age-related diseases. *Curr Genom.* February 2014;15(1):38–51. https://doi.org/10.2174/1389202915011403061333344.

199. Sun-Wang JL, Ivanova S, Zorzano A. The dialogue between the ubiquitin-proteasome system and autophagy: implications in ageing. *Ageing Res Rev.* December 2020;64:101203. https://doi.org/10.1016/j.arr.2020.101203.

200. Kaushik S, Tasset I, Arias E, et al. Autophagy and the hallmarks of aging. *Ageing Res Rev.* December 2021;72:101468. https://doi.org/10.1016/j.arr.2021.101468.

201. Kocaturk NM, Gozuacik D. Crosstalk between mammalian autophagy and the ubiquitin-proteasome system. *Front Cell Dev Biol.* 2018;6:128. https://doi.org/10.3389/fcell.2018.00128.

202. Gamerdinger M, Hajieva P, Kaya AM, Wolfrum U, Hartl FU, Behl C. Protein quality control during aging involves recruitment of the macroautophagy pathway by BAG3. *EMBO J.* April 8, 2009;28(7):889–901. https://doi.org/10.1038/emboj.2009.29.

203. Koyuncu S, Loureiro R, Lee HJ, Wagle P, Krueger M, Vilchez D. Rewiring of the ubiquitinated proteome determines ageing in C. elegans. *Nature.* August 2021;596(7871):285–290. https://doi.org/10.1038/s41586-021-03781-z.

204. Terman A, Brunk UT. Oxidative stress, accumulation of biological 'garbage', and aging. *Antioxidants Redox Signal.* Jan-Feb 2006;8(1–2):197–204. https://doi.org/10.1089/ars.2006.8.197.

205. Carmona-Gutierrez D, Hughes AL, Madeo F, Ruckenstuhl C. The crucial impact of lysosomes in aging and longevity. *Ageing Res Rev.* December 2016;32:2–12. https://doi.org/10.1016/j.arr.2016.04.009.

206. Nixon RA. The aging lysosome: an essential catalyst for late-onset neurodegenerative diseases. *Biochim Biophys Acta Proteins Proteom.* September 2020;1868(9):140443. https://doi.org/10.1016/j.bbapap.2020.140443.

207. Terman A, Kurz T, Navratil M, Arriaga EA, Brunk UT. Mitochondrial turnover and aging of long-lived postmitotic cells: the mitochondrial-lysosomal axis theory of aging. *Antioxidants Redox Signal.* April 2010;12(4):503–535. https://doi.org/10.1089/ars.2009.2598.

208. Boeynaems S, Alberti S, Fawzi NL, et al. Protein phase separation: a new phase in cell biology. *Trends Cell Biol.* June 2018;28(6):420–435. https://doi.org/10.1016/j.tcb.2018.02.004.

209. Franzmann TM, Alberti S. Protein phase separation as a stress survival strategy. *Cold Spring Harbor Perspect Biol.* June 3, 2019;11(6). https://doi.org/10.1101/cshperspect.a034058.

210. Elbaum-Garfinkle S. Matter over mind: liquid phase separation and neuro-degeneration. *J Biol Chem.* May 3, 2019;294(18):7160–7168. https://doi.org/10.1074/jbc.REV118.001188.

211. Oling D, Eisele F, Kvint K, Nystrom T. Opposing roles of Ubp3-dependent deubi-quitination regulate replicative life span and heat resistance. *EMBO J.* April 1, 2014;33(7):747–761. https://doi.org/10.1002/embj.201386822.

212. Weisberg SJ, Lyakhovetsky R, Werdiger AC, Gitler AD, Soen Y, Kaganovich D. Compartmentalization of superoxide dismutase 1 (SOD1G93A) aggregates determines their toxicity. *Proc Natl Acad Sci U S A.* September 25, 2012;109(39):15811–15816. https://doi.org/10.1073/pnas.1205829109.

213. Hachiya N, Sochocka M, Brzecka A, et al. Nuclear envelope and nuclear pore complexes in neurodegenerative diseases-new perspectives for therapeutic interventions. *Mol Neurobiol.* March 2021;58(3):983–995. https://doi.org/10.1007/s12035-020-02168-x.

214. Lionaki E, Gkikas I, Tavernarakis N. Differential protein distribution between the nucleus and mitochondria: implications in aging. *Front Genet.* 2016;7:162. https://doi.org/10.3389/fgene.2016.00162.

215. Gamarra M, de la Cruz A, Blanco-Urrejola M, Baleriola J. Local translation in nervous system pathologies. *Front Integr Neurosci.* 2021;15:689208. https://doi.org/10.3389/fnint.2021.689208.

216. Janssens GE, Meinema AC, Gonzalez J, et al. Protein biogenesis machinery is a driver of replicative aging in yeast. *Elife.* December 1, 2015;4:e08527. https://doi.org/10.7554/eLife.08527.

217. Bergendahl LT, Gerasimavicius L, Miles J, et al. The role of protein complexes in human genetic disease. *Protein Sci.* August 2019;28(8):1400−1411. https://doi.org/10.1002/pro.3667.

218. Toyama BH, Savas JN, Park SK, et al. Identification of long-lived proteins reveals exceptional stability of essential cellular structures. *Cell.* August 29, 2013;154(5):971−982. https://doi.org/10.1016/j.cell.2013.07.037.

219. Krishna S, Arrojo EDR, Capitanio JS, Ramachandra R, Ellisman M, Hetzer MW. Identification of long-lived proteins in the mitochondria reveals increased stability of the electron transport chain. *Dev Cell.* November 8, 2021;56(21):2952−2965 e9. https://doi.org/10.1016/j.devcel.2021.10.008.

220. Rempel IL, Crane MM, Thaller DJ, et al. Age-dependent deterioration of nuclear pore assembly in mitotic cells decreases transport dynamics. *Elife.* June 3, 2019;8. https://doi.org/10.7554/eLife.48186.

221. Rempel IL, Steen A, Veenhoff LM. Poor old pores-The challenge of making and maintaining nuclear pore complexes in aging. *FEBS J.* March 2020;287(6):1058−1075. https://doi.org/10.1111/febs.15205.

222. Gomez LA, Hagen TM. Age-related decline in mitochondrial bioenergetics: does supercomplex destabilization determine lower oxidative capacity and higher superoxide production? *Semin Cell Dev Biol.* September 2012;23(7):758−767. https://doi.org/10.1016/j.semcdb.2012.04.002.

223. Amirbeigiarab S, Kiani P, Velazquez Sanchez A, et al. Invariable stoichiometry of ribosomal proteins in mouse brain tissues with aging. *Proc Natl Acad Sci U S A.* November 5, 2019;116(45):22567−22572. https://doi.org/10.1073/pnas.1912060116.

224. Dues DJ, Andrews EK, Schaar CE, Bergsma AL, Senchuk MM, Van Raamsdonk JM. Aging causes decreased resistance to multiple stresses and a failure to activate specific stress response pathways. *Aging.* April 2016;8(4):777−795. https://doi.org/10.18632/aging.100939.

225. Vermeulen CJ, Loeschcke V. Longevity and the stress response in Drosophila. *Exp Gerontol.* March 2007;42(3):153−159. https://doi.org/10.1016/j.exger.2006.09.014.

226. Colinet H, Chertemps T, Boulogne I, Siaussat D. Age-related decline of abiotic stress tolerance in young Drosophila melanogaster adults. *J Gerontol A Biol Sci Med Sci.* December 2016;71(12):1574−1580. https://doi.org/10.1093/gerona/glv193.

227. Ukraintseva S, Arbeev K, Duan M, et al. Decline in biological resilience as key manifestation of aging: potential mechanisms and role in health and longevity. *Mech Ageing Dev.* March 2021;194:111418. https://doi.org/10.1016/j.mad.2020.111418.

228. Ottens F, Franz A, Hoppe T. Build-UPS and break-downs: metabolism impacts on proteostasis and aging. *Cell Death Differ.* February 2021;28(2):505−521. https://doi.org/10.1038/s41418-020-00682-y.

229. Pan H, Finkel T. Key proteins and pathways that regulate lifespan. *J Biol Chem.* April 21, 2017;292(16):6452−6460. https://doi.org/10.1074/jbc.R116.771915.

230. Wang T, Wang W, Wang Q, Xie R, Landay A, Chen D. The E3 ubiquitin ligase CHIP in normal cell function and in disease conditions. *Ann N Y Acad Sci.* January 2020;1460(1):3−10. https://doi.org/10.1111/nyas.14206.

231. Tawo R, Pokrzywa W, Kevei E, et al. The ubiquitin ligase CHIP integrates proteostasis and aging by regulation of insulin receptor turnover. *Cell*. April 20, 2017;169(3):470–482 e13. https://doi.org/10.1016/j.cell.2017.04.003.

232. Mathew R, Pal Bhadra M, Bhadra U. Insulin/insulin-like growth factor-1 signalling (IIS) based regulation of lifespan across species. *Biogerontology*. February 2017;18(1):35–53. https://doi.org/10.1007/s10522-016-9670-8.

233. Lee JY, Davis I, Youth EHH, et al. Misexpression of genes lacking CpG islands drives degenerative changes during aging. *Sci Adv*. December 17, 2021;7(51):eabj9111. https://doi.org/10.1126/sciadv.abj9111.

234. Li C, Casanueva O. Epigenetic inheritance of proteostasis and ageing. *Essays Biochem*. October 15, 2016;60(2):191–202. https://doi.org/10.1042/EBC20160025.

235. Echtenkamp FJ, Gvozdenov Z, Adkins NL, et al. Hsp90 and p23 molecular chaperones control chromatin architecture by maintaining the functional pool of the RSC chromatin remodeler. *Mol Cell*. December 1, 2016;64(5):888–899. https://doi.org/10.1016/j.molcel.2016.09.040.

236. Benayoun BA, Pollina EA, Singh PP, et al. Remodeling of epigenome and transcriptome landscapes with aging in mice reveals widespread induction of inflammatory responses. *Genome Res*. April 2019;29(4):697–709. https://doi.org/10.1101/gr.240093.118.

237. DeZwaan DC, Freeman BC. HSP90 manages the ends. *Trends Biochem Sci*. July 2010;35(7):384–391. https://doi.org/10.1016/j.tibs.2010.02.005.

238. Freund A, Zhong FL, Venteicher AS, et al. Proteostatic control of telomerase function through TRiC-mediated folding of TCAB1. *Cell*. December 4, 2014;159(6):1389–1403. https://doi.org/10.1016/j.cell.2014.10.059.

239. Chakravarti D, LaBella KA, DePinho RA. Telomeres: history, health, and hallmarks of aging. *Cell*. January 21, 2021;184(2):306–322. https://doi.org/10.1016/j.cell.2020.12.028.

240. Gumeni S, Evangelakou Z, Gorgoulis VG, Trougakos IP. Proteome stability as a key factor of genome integrity. *Int J Mol Sci*. September 22, 2017;18(10). https://doi.org/10.3390/ijms18102036.

241. Ainslie A, Huiting W, Barazzuol L, Bergink S. Genome instability and loss of protein homeostasis: converging paths to neurodegeneration? *Open Biol*. April 2021;11(4):200296. https://doi.org/10.1098/rsob.200296.

242. Andreasson C, Ott M, Buttner S. Mitochondria orchestrate proteostatic and metabolic stress responses. *EMBO Rep*. October 4, 2019;20(10):e47865. https://doi.org/10.15252/embr.201947865.

243. Lu B, Guo S. Mechanisms linking mitochondrial dysfunction and proteostasis failure. *Trends Cell Biol*. April 2020;30(4):317–328. https://doi.org/10.1016/j.tcb.2020.01.008.

244. Zimmermann A, Madreiter-Sokolowski C, Stryeck S, Abdellatif M. Targeting the mitochondria-proteostasis Axis to delay aging. *Front Cell Dev Biol*. 2021;9:656201. https://doi.org/10.3389/fcell.2021.656201.

245. Su KH, Dai C. mTORC1 senses stresses: coupling stress to proteostasis. *Bioessays*. May 2017;39(5). https://doi.org/10.1002/bies.201600268.

246. Zhu X, Chen Z, Shen W, et al. Inflammation, epigenetics, and metabolism converge to cell senescence and ageing: the regulation and intervention. *Signal Transduct Targeted Ther*. June 28, 2021;6(1):245. https://doi.org/10.1038/s41392-021-00646-9.

247. Abbadie C, Pluquet O. Unfolded protein response (UPR) controls major senescence hallmarks. *Trends Biochem Sci*. May 2020;45(5):371–374. https://doi.org/10.1016/j.tibs.2020.02.005.

248. Sabath N, Levy-Adam F, Younis A, et al. Cellular proteostasis decline in human senescence. *Proc Natl Acad Sci U S A*. December 15, 2020;117(50):31902–31913. https://doi.org/10.1073/pnas.2018138117.

249. Cavinato M, Madreiter-Sokolowski CT, Buttner S, et al. Targeting cellular senescence based on interorganelle communication, multilevel proteostasis, and metabolic control. *FEBS J*. June 2021;288(12):3834–3854. https://doi.org/10.1111/febs.15631.

250. Vilchez D, Simic MS, Dillin A. Proteostasis and aging of stem cells. *Trends Cell Biol.* March 2014;24(3):161—170. https://doi.org/10.1016/j.tcb.2013.09.002.

251. Leeman DS, Hebestreit K, Ruetz T, et al. Lysosome activation clears aggregates and enhances quiescent neural stem cell activation during aging. *Science.* March 16, 2018;359(6381):1277—1283. https://doi.org/10.1126/science.aag3048.

252. Schuler SC, Gebert N, Ori A. Stem cell aging: the upcoming era of proteins and metabolites. *Mech Ageing Dev.* September 2020;190:111288. https://doi.org/10.1016/j.mad.2020.111288.

253. Taylor RC, Berendzen KM, Dillin A. Systemic stress signalling: understanding the cell non-autonomous control of proteostasis. *Nat Rev Mol Cell Biol.* March 2014;15(3):211—217. https://doi.org/10.1038/nrm3752.

254. Guix FX. The interplay between aging-associated loss of protein homeostasis and extracellular vesicles in neurodegeneration. *J Neurosci Res.* February 2020;98(2):262—283. https://doi.org/10.1002/jnr.24526.

255. Moehle EA, Shen K, Dillin A. Mitochondrial proteostasis in the context of cellular and organismal health and aging. *J Biol Chem.* April 5, 2019;294(14):5396—5407. https://doi.org/10.1074/jbc.TM117.000893.

256. Bhattarai KR, Riaz TA, Kim HR, Chae HJ. The aftermath of the interplay between the endoplasmic reticulum stress response and redox signaling. *Exp Mol Med.* February 2021;53(2):151—167. https://doi.org/10.1038/s12276-021-00560-8.

257. Matos L, Gouveia AM, Almeida H. ER stress response in human cellular models of senescence. *J Gerontol A Biol Sci Med Sci.* August 2015;70(8):924—935. https://doi.org/10.1093/gerona/glu129.

258. Ishikawa S, Ishikawa F. Proteostasis failure and cellular senescence in long-term cultured postmitotic rat neurons. *Aging Cell.* January 2020;19(1):e13071. https://doi.org/10.1111/acel.13071.

259. Omer A, Patel D, Moran JL, Lian XJ, Di Marco S, Gallouzi IE. Autophagy and heat-shock response impair stress granule assembly during cellular senescence. *Mech Ageing Dev.* December 2020;192:111382. https://doi.org/10.1016/j.mad.2020.111382.

260. Chondrogianni N, Gonos ES. Proteasome inhibition induces a senescence-like phenotype in primary human fibroblasts cultures. *Biogerontology.* 2004;5(1):55—61. https://doi.org/10.1023/b:bgen.0000017687.55667.42.

261. Bernard M, Yang B, Migneault F, et al. Autophagy drives fibroblast senescence through MTORC2 regulation. *Autophagy.* January 13, 2020:1—13. https://doi.org/10.1080/15548627.2020.1713640.

262. Llamas E, Alirzayeva H, Loureiro R, Vilchez D. The intrinsic proteostasis network of stem cells. *Curr Opin Cell Biol.* December 2020;67:46—55. https://doi.org/10.1016/j.ceb.2020.08.005.

263. Yan P, Ren J, Zhang W, Qu J, Liu GH. Protein quality control of cell stemness. *Cell Regen.* November 12, 2020;9(1):22. https://doi.org/10.1186/s13619-020-00064-2.

264. Taylor RC, Hetz C. Mastering organismal aging through the endoplasmic reticulum proteostasis network. *Aging Cell.* November 2020;19(11):e13265. https://doi.org/10.1111/acel.13265.

265. Morimoto RI. Cell-nonautonomous regulation of proteostasis in aging and disease. *Cold Spring Harbor Perspect Biol.* April 1, 2020;12(4). https://doi.org/10.1101/cshperspect.a034074.

266. Kulkarni AS, Gubbi S, Barzilai N. Benefits of metformin in attenuating the hallmarks of aging. *Cell Metabol.* July 7, 2020;32(1):15—30. https://doi.org/10.1016/j.cmet.2020.04.001.

267. Johnson SC, Rabinovitch PS, Kaeberlein M. mTOR is a key modulator of ageing and age-related disease. *Nature.* January 17, 2013;493(7432):338—345. https://doi.org/10.1038/nature11861.

268. Gonzalez-Freire M, Diaz-Ruiz A, Hauser D, et al. The road ahead for health and lifespan interventions. *Ageing Res Rev*. May 2020;59:101037. https://doi.org/10.1016/j.arr.2020.101037.

269. Derisbourg MJ, Hartman MD, Denzel MS. Modulating the integrated stress response to slow aging and ameliorate age-related pathology. *Nature Aging*. 2021;1(9):760−768. https://doi.org/10.1038/s43587-021-00112-9, 2021/09/01.

270. Martinez-Miguel VE, Lujan C, Espie-Caullet T, et al. Increased fidelity of protein synthesis extends lifespan. *Cell Metabol*. November 2, 2021;33(11):2288−2300 e12. https://doi.org/10.1016/j.cmet.2021.08.017.

271. Tain LS, Sehlke R, Jain C, et al. A proteomic atlas of insulin signalling reveals tissue-specific mechanisms of longevity assurance. *Mol Syst Biol*. September 15, 2017;13(9):939. https://doi.org/10.15252/msb.20177663.

272. Williams R, Laskovs M, Williams RI, Mahadevan A, Labbadia J. A mitochondrial stress-specific form of HSF1 protects against age-related proteostasis collapse. *Dev Cell*. September 28, 2020;54(6):758−772 e5. https://doi.org/10.1016/j.devcel.2020.06.038.

273. Fuhrmann-Stroissnigg H, Ling YY, Zhao J, et al. Identification of HSP90 inhibitors as a novel class of senolytics. *Nat Commun*. September 4, 2017;8(1):422. https://doi.org/10.1038/s41467-017-00314-z.

274. Kucheryavenko O, Nelson G, von Zglinicki T, Korolchuk VI, Carroll B. The mTORC1-autophagy pathway is a target for senescent cell elimination. *Biogerontology*. June 2019;20(3):331−335. https://doi.org/10.1007/s10522-019-09802-9.

275. Limbad C, Doi R, McGirr J, et al. Senolysis induced by 25-hydroxycholesterol targets CRYAB in multiple cell types. *iScience*. 2022;25(2):103848. https://doi.org/10.1016/j.isci.2022.103848, 2022/02/18/.

276. Manasanch EE, Orlowski RZ. Proteasome inhibitors in cancer therapy. *Nat Rev Clin Oncol*. July 2017;14(7):417−433. https://doi.org/10.1038/nrclinonc.2016.206.

277. Workman P. Reflections and outlook on targeting HSP90, HSP70 and HSF1 in cancer: a personal perspective. *Adv Exp Med Biol*. 2020;1243:163−179. https://doi.org/10.1007/978-3-030-40204-4_11.

278. Janssens GE, Lin XX, Millan-Arino L, et al. Transcriptomics-based screening identifies pharmacological inhibition of Hsp90 as a means to defer aging. *Cell Rep*. April 9, 2019;27(2):467−480 e6. https://doi.org/10.1016/j.celrep.2019.03.044.

279. Gems D, Partridge L. Stress-response hormesis and aging: "that which does not kill us makes us stronger". *Cell Metabol*. March 2008;7(3):200−203. https://doi.org/10.1016/j.cmet.2008.01.001.

SECTION III

Autophagy and bioenergetics

CHAPTER 5

Autophagy and bioenergetics in aging

Jianying Zhang[1,2], He-Ling Wang[1] and Evandro Fei Fang[1,3]

[1]Department of Clinical Molecular Biology, University of Oslo and Akershus University Hospital, Lørenskog, Norway; [2]Xiangya School of Stomatology, Central South University, Changsha, China; [3]The Norwegian Centre on Healthy Ageing (NO-Age) Network, Oslo, Norway

1. Compromised autophagy and mitophagy in aging and disease

1.1 Autophagy, mitophagy, and aging

Aging is a biological process characterized by a progressive decline in cellular and physiologic functions, resulting in the reduced quality of life for the organism involved.[1] Accordingly, aging is a primary risk factor for developing multiple disorders, including cardiovascular disease (e.g., stroke), metabolic syndrome (e.g., type 2 diabetes), cancer, and neurodegenerative disease such as Alzheimer's disease (AD). As a whole, age-related ailments pose a formidable socioeconomic burden and present a substantial health care challenge to the global community.[2,3] It is therefore of paramount importance to identify therapeutic interventions that can promote healthy aging, the maintenance of functional ability in old age by which the elderly can perform daily activities on their own, while halting the development of multiple age-related pathologic conditions.[4] Among the numerous molecular changes associated with aging, alteration of autophagy has been reported across diverse species as integral to aging. Autophagy describes different routes that cells apply to recognize and deliver cytoplasmic and nuclear substrates to lysosomes for degradation.[5] The development of deeper insight into the temporal and spatial effects of impaired autophagy on tissue homeostasis over the past few decades has revealed a complex and multifaceted relationship between autophagy and aging. In this chapter, we analyze the relationships among autophagy, mitophagy, aging, and age-related disease, and attempt to understand how decreases in the function of specific phagosome systems, including autophagosomes and mitochondria-dedicated mitophagosomes, can affect recycling of dysfunctional cellular components, exacerbate aging, and ultimately negatively influence long-term tissue function and organismal health.

Molecular, Cellular, and Metabolic Fundamentals of Human Aging
ISBN 978-0-323-91617-2
https://doi.org/10.1016/B978-0-323-91617-2.00002-X

© 2023 Elsevier Inc.
All rights reserved.

A number of studies over the past decades have shown that the process of autophagy can take many different forms. As a highly conserved pathway, autophagy is responsible for degrading cellular components via lysosomes, such as defective organelles and aggregates of misfolded proteins. The concept of autophagy was first described in the 1960s. In the 1990s, however, the discovery of autophagy-related genes (ATGs) was the catalyst for major advances in the underlying molecular mechanisms.[6–9] Generally, there are three primary types of autophagy: macroautophagy, micro-autophagy, and chaperone-mediated autophagy (CMA), all of which involve the delivery of substrates to the lysosome for degradation. Macroautophagy involves the formation of double-membrane vesicles, called autophagosomes, to sequester and deliver cargoes to the lysosome. The process of microautophagy is characterized by the direct uptake of cargoes through invagination of the lysosomal membrane. While in CMA, individual unfolded proteins are transported directly across the lysosomal membrane.[10] In the beginning, macroautophagy (here referred to as autophagy) was thought to be a nonselective, bulk degradation process. Nevertheless, the discovery of the first mammalian selective autophagy receptor in 2007, p62/SQSTM1, changed this belief.[11] p62/SQSTM1, as well as subsequently identified proteins including neighbor of BRCA1 gene 1 (NBR1), nuclear domain 10 protein 52 (NDP52), optineurin (OPTN), and tax1-bing protein 1, has defined a new family of autophagy-related proteins. Those proteins function to target protein aggregates, mitochondria, intracellular pathogens, and other cargoes to the core autophagy machinery through an LC3-interacting region motif.[12] Today, autophagy is recognized as a highly selective cellular clearance pathway essential for maintaining cellular and tissue homeostasis. Based on the specific subcellular cargoes involved, selective autophagy can be further classified into several subtypes. These include varieties that specialize in macromolecules (glycophagy and lipophagy), mitochondria (mitophagy), endoplasmic reticulum (ER) (reticulophagy), parts of the nucleus (nucleophagy), pathogenic infection (xenophagy), and the lysosomes themselves (lysophagy). Table 5.1 gives a general overview of the differences in selected cargoes and corresponding receptors and adaptors among the subtypes.[13–18] The core process of macroautophagy has been described in greater detail elsewhere.[14] Autophagy could be initiated by inhibition of mammalian target of rapamycin (mTOR) or activation of 5′AMP-activated protein kinase (AMPK), which are canonical inducers of autophagy after various physiologic stresses (e.g., starvation, high temperature) and physical exercise. Transcription

Table 5.1 Selected cargoes and corresponding mammalian receptors and adaptors in selective autophagy.

Selective type of autophagy	Selected cargoes	Receptors and adaptors in mammals
Glycophagy	Glycogen	Starch-binding domain-containing protein 1
Lipophagy	Lipid droplets	p62/sequestosome 1, adipose triglyceride lipase, patatin-like phospholipase domain containing 8
Mitophagy	Mitochondria	Autophagy and Beclin 1 regulator 1, BCL2-like, BCL2 interacting protein 3, cardiolipin, ceramide, FKBP prolyl isomerase 8, FUN14 domain-containing protein 1, calcium binding and coiled-coil domain 2/NDP52, NIX, nucleotide-binding, leucine-rich repeat containing X1, optineurin, p62/sequestosome 1, prohibitin 2, tax1-binding protein 1, neighbor of BRCA1 gene 1, mitochondrial Rho GTPase 1
Reticulophagy	Endoplasmic reticulum	Atlastin 3, C53, cell cycle progression gene 1, FAM134B, reticulon 3, SEC62, testis expressed 264, calcium binding and coiled-coil domain 1, tripartite motif containing 13
Nucleophagy	Nucleus fragments	—
Xenophagy	Cellular pathogens	NDP52, optineurin, p62/sequestosome 1, tax1-binding protein 1, neighbor of BRCA1 gene 1
Lysophagy	Lysosomes	p62/sequestosome 1, tripartite motif containing 16

factor EB (TFEB) is another important autophagy and lysosomal biogenesis promoter whose nuclear translocation is coordinated by both mTOR (via phosphorylation) and AMPK activity (via folliculin). Upon activation, the autophagic process begins with the nucleation of membranes and formation of phagophores, which then undergo elongation and maturation before the autophagosome fuses with the lysosome for cargo degradation and recycling (the core machinery of autophagy is demonstrated in Fig. 5.1; Table 5.2 provides an overview of autophagy-related orthologue genes across species). Several studies suggested that membranes derived from the Golgi

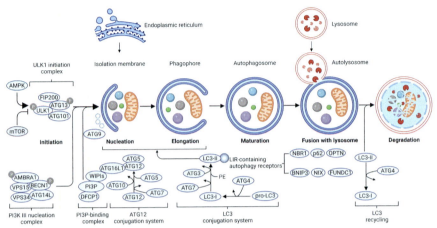

Figure 5.1 *Core machinery of macroautophagy. The main regulatory machinery of macroautophagy (autophagy) is shown. Autophagy is a multistep process that includes (1) initiation, (2) membrane nucleation, (3) phagophore formation and elongation, (4) autophagosome maturation, (5) fusion with the lysosome, and (6) degradation, which are regulated by multiple proteins, referred to as autophagy-related proteins (ATGs). The initiation of autophagy requires the ULK1 kinase complex, which is tightly regulated by AMP-activated kinase (AMPK) and mammalian target of rapamycin (mTOR), with AMPK acting as an activator while mTOR acts as an inhibitor. When autophagy is induced, AMPK activates ULK1 through phosphorylation. The ULK1 initiation complex (composed of ULK1, FIP200, ATG13, and ATG101) stimulates the class III phosphatidylinositol 3-kinase complex (PI3K III nucleation complex), which includes BECN1, AMBRA1, ATG14L, VPS15, and VPS34. This complex then produces a pool of PtdIns3P, which leads to the recruitment of WD-repeat protein interacting with phosphoinositides (WIPI) proteins, which will recover ATG-9 vesicles from previous membranes and recruit ATG5-ATG12-ATG16L1 (E3). The ATG3—ATG12—ATG16L1 complex then promotes conjugation of LC3, in which LC3 is cleaved by the protease ATG4 to form LC3-I, which is further recognized by E1 (ATG7), E2 (ATG3), and E3 and then conjugated with phosphatidylethanolamine (PE) to form LC3-II. LC3-II binds to the LC3-interacting region (LIR)-containing autophagy receptors that bind to items targeted for degradation. For simplicity, only the names of vertebrate ATGs are shown here.*

complex, mitochondria, and the plasma membrane also contribute to the formation of autophagosomes. These findings suggest that autophagosome formation involves multiple interdependent processes.[19] The mechanisms involved in selective autophagy may differ according to the cargo, but they are mostly the same as those in nonselective autophagy. Selectivity arises because certain autophagy proteins or other effectors can specifically recognize receptor molecules on the surface of cellular components such as organelles, pathogens, or large protein aggregates, and initiate the formation

Table 5.2 Autophagy-related genes in yeasts, worms, flies, and mammals.

Autophagy-related orthologue genes across species					Function in autophagy
Saccharomyces cerevisiae	Caenorhabditis elegans	Drosophila melanogaster	Human (Homo sapiens)	Mouse (Mus musculus)	
Regulation of autophagy induction (inexhaustive)					
TOR1/ TOR2	let-363; B0261.2	Tor	mTOR	Mtor	Rapamycin–sensitive ser/Thr protein kinase
ATG1	unc-51	Atg1	ULK1	Ulk1	Ser/Thr protein kinase
ATG13	epg-1; D2007.5	Atg13	ATG13	Atg13	Phosphoprotein component of Atg1 complex
Vesicle nucleation					
ATG6; VPS30	bec-1; T19E7.3	Atg6	BECN1	Becn1	Component of class III PI3-kinase complex
VPS34	let-512; B0025.1	Pi3K59F	PIK3C3	Pik3c3	Kinase that produces PI3P to enable recruitment of machinery that forms autophagosomes
VPS15	Vps15 ZK930.1	Vps15	PIK3R4	Pik3r4	Component of class III PI3-kinase complex
Vesicle expansion and completion					
ATG3	atg-3; Y55F3AM.4	Atg3	ATG3	Atg3	E2-like enzyme; conjugates Atg8 to PE
ATG4	atg-4; Y87G2A.3	Atg4	ATG4	Atg4	Cysteine protease; leaves C-terminal extension of PE from Atg8
ATG5	atg-5; Y71G12B.12	Atg5	ATG5	Atg5	Conjugated to Atg12 through internal lysine
ATG7		Atg7	ATG7	Atg7	

(Continued)

	atgr-7; M7.5				E1-like enzyme for Atg5-Atg 12 complex and Atg8 conjugation systems
ATG8	lgg-1; C35D5.9	Atg8a/8b	GABARAP	Gabarap	Ubiquitin-like protein conjugated to PE
ATG10	atg-10; D2085.2	Atg10	ATG10	Atg10	E2-like enzyme; conjugates Atg12 to Atg5
ATG12	lgg-3; B0336.8	Atg12	APG12	Atg12	Ubiquitin-like protein conjugated to Atg5
ATG16	atg-16.2 (K06A1.5); atg-16.1 (F02E8.5)	Atg16	ATG16L1/L2	Atg16l1/l2	Component of Atg12–Atg5 complex
Recycling of autophagic proteins					
ATG2	atg-2; M03A8.2	Atg2	ATG2A/2B	Atg2a/2b	Peripheral membrane protein; interacts with Atg9
ATG9	atgr-9; T22H9.2	Atg9	ATG9A/9B	Atg9a/9b	Integral membrane protein; interacts with Atg2
ATG18	atg-18; F41E6.13	Atg18a/18b	WIPI1/2	Wipi1/2	Peripheral membrane protein; binds PI3P and PI (3,5) P; localization of Atg2

ATG, autophagy-related gene; GABARAP, γ-aminobutyric acid receptor-associated protein; mTOR, mechanistic (or mammalian) target of rapamycin kinase; PIK3C3, phosphatidylinositol 3-kinase catalytic subunit type 3; PIK3R4, phosphoinositide-3-kinase regulatory subunit 4; ULK1, unc-51 like autophagy activating kinase 1; VPS, vacuolar protein sorting; WIPI1/2, WD, repeat domain, phosphoinositide interacting 1/2.
Genes are listed according to their functions in each step of autophagy.

of autophagosomes. Moreover, autophagy receptors and specific adapters such as p62/SQSTM1 and NBR1 are not required for nonselective autophagy.[20,21]

During aging, damage to subcellular organelles accumulates. To maintain cellular functionality and viability, dysfunctional organelles must be disposed of and recycled promptly and efficiently. Among the various types of autophagy able to target different subcellular organelles, mitophagy has perhaps been the most extensively studied.[22] Mitophagy is the process of selectively eliminating defective or surplus mitochondria through autophagy (an overview of mitophagy pathway is shown in Fig. 5.2). The PTEN-induced putative protein kinase 1 (PINK1)-Parkin mediated pathway for degradation of heavily depolarized mitochondria is the most well-understood pathway.[10] Specifically, part of this process involves Ser-65 phosphorylated ubiquitin attracting soluble autophagy receptors NDP52, OPTN, and p62/SQSTM1, which then recruit the core components of autophagy to form the mitophagosome around the damaged mitochondrion. Furthermore, other mitophagy pathways (basal, developmental, and stress-induced) involve binding LC3 to a series of mitochondrial outer membrane proteins that contain LC3-interacting region domains such as NIX (BNIP3-like [BNIP3L]), BNIP3, FKBP8, FUNDC1, BCL2L13, PHB2, and AMBRA1, and Atg8/LC3-binding mitochondrial lipids such as cardiolipin.[23] Imbalances in the quality surveillance system of mitochondria may contribute to the development of age-related pathologies and premature aging processes.[24] Further studies on mitophagy are essential to decipher their multilayered regulatory network, as well as their relationship to aging and health. A specific focus of these studies would include works to understand how these processes change with aging, and how these changes influence age-related tissue function, which will eventually lead to insights concerning human health and quality of life.

1.2 Tissue-specific autophagy and mitophagy in aging

Because aging is associated with functional decline at both the tissue and organismal levels, it is important to understand how aging within the individual tissues affects and is affected by aging across the entire organism.[25] The role of autophagy in regulating aging appears to be tissue-specific, as evidenced by studies with nematodes (*Caenorhabditis elegans*), fruit flies (*Drosophila melanogaster*), and mice.[26–28] In nematodes, two long-lived mutant strains, dietary-restricted *eat-2* and insulin growth factor 1 (IGF-1)

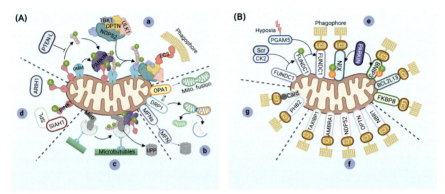

Figure 5.2 *Overview of mitophagy pathways.* (A) The PTEN-induced putative kinase 1 (PINK1)−parkin pathway of mitophagy. (a) For the PINK1- and parkin-dependent mitophagy pathway, reduced mitochondrial membrane potential (ψm) enables the stabilization of full-length PINK1 at the outer mitochondrial membrane (OMM), presumably through the OMM complex (TOM complex). Full-length autophosphorylated PINK1 phosphorylates ubiquitin enables the binding of cytosolic parkin to phosphoubiquitin and the recruitment and binding of OMM receptors. Parkin, phosphorylated by PINK1, ubiquitinates several known receptors, including voltage-dependent anion-selective channel 1 (VDAC1), mitofusins (MFN1 and MFN2), and mitochondria Rho-GTPase (Miro). (b) The prevention of damaged mitochondrial fusion from the healthy mitochondrial network, via parkin, MFN1/2, optic atrophy 1 (OPA1), and Drp1, facilitates mitophagy, because smaller mitochondria can easily be sequestered, engulfed, and degraded. (c) PINK1 and parkin arrest mitochondrial motility by degrading Miro, quarantining damaged mitochondria, and promoting their mitophagosomal engulfment. (d) PINK1-dependent, parkin-independent mitophagy pathway. In addition to parkin, alternative E3 ubiquitin ligases such as SIAH1 and ARIH1 participate in PINK1-dependent, parkin-independent mitophagy. (B) Receptor-mediated mitophagy pathways. Receptor-mediated mitophagy relies on various proteins that can be divided into three main groups. (e) Group 1 is composed of OMM proteins, including FUN14 domain-containing protein 1 (FUNDC1), BCL2 interacting protein 3 (BNIP3), Nip3-like protein X (NIX, also named BNIP3L), BCL2-like 13 (BCL2L13), and FKBP prolyl isomerase 8 (FKBP8). (f) Group 2 encompasses numerous autophagy adaptors involved in mitophagy, including BRCA1 gene 1 (NBR1), optineurin (OPTN), nuclear domain 10 protein 52 (NDP52), autophagy and Beclin 1 regulator 1 (AMBRA1), and TAX1 binding protein 1 (TAX1BP1). (g) Group 3 contains mitophagy receptors that are inner mitochondrial membrane (IMM) components, including PHB2 and cardiolipin (Card).

receptor-deficient *daf-2*, could be reduced in life span by inhibiting *lgg-1* and *atg-18* specifically in body-wall muscles of adults.[29,30] Furthermore, ATG 18 is a member of the WD-repeat protein interacting with phosphoinositides (WIPI) protein family, homologous to mammalian WIPI1 and WIPI2. Restoring the function of ATG-18 in particular tissues may make it possible to suppress the shortened life span of *atg-18* mutant *C. elegans*.[30] For example, pan-neuronal or intestinal overexpression of ATG-18 results in the complete restoration of the life span of mutant *atg-18* worms to that of wild-type nematodes, whereas muscle or hypodermis-targeted recovery of ATG-18 had little effect on life span restoration.[31] In flies, enhancement of autophagy in muscle tissue through overexpression of Atg-8a or Forkhead box O (FOXO) was sufficient for the extension of life span,[32] whereas in mice, a deficiency of muscle-specific ATG-7 led to impaired muscle function (possibly via mitochondrial dysfunction) and decreased life span as a result of the inhibition of autophagy.[33] Moreover, enhancing intestinal autophagy improved intestinal function and increased health span and life span in both nematodes and flies.[29] Because our tissues age unevenly, resulting in faster degeneration in some tissues than in others, it would be interesting to determine how closely rates of aging and autophagy are correlated in various tissues throughout an organism's life span.[34]

1.3 Compromised autophagy or mitophagy is a hallmark of aging

An increasing body of evidence suggests that autophagic activity declines with age in diverse organisms. Studies in nematodes, rodents, and in vitro human cells demonstrated that the proteolytic function of lysosomes declines with age, leading to a reduction in autophagic flux and the aggravation of age-related disease.[35–37] Further evidence from flies shows that aging is associated with reduced expression of several *Atg* genes (e.g., *Atg2*, *Atg-8a*) (LC3/GABARAP in mammals), which are pivotal for initiating and promoting autophagy.[38] Autophagy is impaired in the hypothalamus of aged wild-type mice, as evidenced by reduced rates of autophagolysosomal fusion and impaired delivery of autophagic cargoes to lysosomes.[39] In addition, autophagic processes were reduced in 18- to 25-month-old murine brain tissue, as evidenced by reductions in *Atg5-Atg12* conjugate and *Becn1* levels, elevated mTOR activity, and increased ferritin H levels.[40] Emerging findings in aged rats pointed to a decrease in the expression of autophagy-related protein Becn1, which is important for microglia-dependent phagocytosis in whole brain tissue, as well as in the hippocampus of naked mole rats and Wistar rats, respectively.[41,42] Finally, compromised autophagy is one of the most common factors contributing to the collapse of tissue homeostasis in elderly mice and in rodent models of human disease.[43] Specifically, age-associated dysregulation of autophagy, which manifests as the accumulation of autophagosomes, possibly owing to impaired lysosomal fusion and degradation, contributes to neurodegeneration as well as cardiac and skeletal muscle aging.[44] The results of human studies agree with observations in rodent models, in which expression levels of genes related to autophagy, such as *ATG5*, *ATG7*, and *BECN1*, decline with age. As a result, age-dependent declines in autophagy are closely associated with the development and progression of several human pathologies.[45] These studies show that cargo delivery to lysosomes and expression of autophagy-associated proteins gradually decline with age, indicating that impaired autophagy might be a key factor involved in aging in living organisms.

The accumulation of dysfunctional mitochondria is a hallmark of aging. Although the mechanisms that underly age-related loss of mitochondrial function remain incompletely understood and may involve a variety of processes, it has been proposed that a decline in mitophagy is an important factor in aging.[6] The degradation of damaged mitochondria occurs in mammals through a pathway consisting of the mitochondrial serine/threonine-protein kinase PINK1 and the E3 ubiquitin-protein ligase Parkin. Studies of mammalian cells in culture and genetic studies on model organisms have provided insights into the molecular mechanisms of mitophagy.[46–48] Several age-related conditions, including heart disease, retinopathy, fatty liver disease, pulmonary hypertension, kidney disease, Parkinson's disease (PD), amyotrophic lateral sclerosis (ALS), AD, and Huntington's disease (HD), are all linked to disruptions in mitophagy.[49–51] Age-related declines in mitophagy markers have been observed in several model organisms, including nematodes and flies. In *C. elegans*, DCT-1 (DAF-16/FOXO controlled germ line tumor affecting-1, an NIP3 homolog) is a putative orthologue of BNIP3 and BNIP3L proteins, which are mitophagy receptors. An increase in mitochondrial content is observed upon inhibition of DCT-1, which suggests that DCT-1 functions as a critical regulator of mitophagy.[50] In flies, loss of PINK1 or parkin leads to early onset behavioral decline and shortened life span, whereas the ubiquitous or neuron-specific upregulation of Parkin throughout adulthood extends their life span.[52] Furthermore, promoting dynamin-related protein 1 (Drp1)-mediated mitochondrial fission in the midlife of flies restores mitochondrial morphology to a youthful state, facilitates mitophagy, and enhances mitochondrial respiratory function. Transient Drp1 induction during midlife improved markers of organismal health and delayed age-related gut pathology while extending fly life span in a manner attributable to Atg1.[53] In addition, upregulating Drp1 from midlife onward, specifically in neurons or the intestine, prolongs the life span of flies.[53] These results indicate that declines in mitophagy during middle age may partly result from altered mitochondrial dynamics, and that they might lead to mitochondrial dysfunction and shortened life spans, at least in *D. melanogaster*.

In light of these findings, it was suggested that pharmacologic interventions that stimulate mitophagy may prove effective in delaying the health decline associated with aging. Among these interventions are nicotinamide adenine dinucleotide (oxidized form, NAD^+), a fundamental metabolite in energy metabolism, redox homeostasis, mitochondrial function, and the regulation of cell survival and death.[51] NAD^+-activated sirtuins (SIRTs) activate autophagy by inhibiting mTOR and deacetylating several autophagy proteins (ATG5, ATG7, and ATG8). A series of mitophagy-related proteins, including PINK1, parkin, NIX (in the DCT-1 worm), and BNIP3, are activated by the NAD^+—SIRT axis. In nematodes, flies, and mice, NAD^+ precursors, such as nicotinamide, nicotinamide riboside (NR), and nicotinamide mononucleotide (NMN), can increase life span and possibly improve health span.[54] Urolithin A (UA), a metabolite of the gut microflora, is another clinically promising mitophagy inducer. UA extends worm life span by regulating genes involved in autophagy (i.e., *bec-1*, *sqst-1*, *vps-34*) and mitophagy (*pink-1*, *dct-1*, and *Nrf2/skn-1*).[55] UA prevents memory loss in both Aβ and tau AD models in *C. elegans* and mouse in a mitophagy-dependent manner (*pink-1*, *pdr-1*, or *dct-1*).[50] There there have been advances in identifying novel and well-known compounds that induce autophagy, but it remains highly important to highlight the multifaceted effects of these therapeutic agents and to understand how they interact with a full complement of targets so that they can be used reliably for therapeutic interventions.

1.4 Dysfunctional autophagy and mitophagy lead to accelerated aging

As an important intracellular quality control system in removing damaged or long-lived organelles and protein aggregates, autophagy impairment exacerbates aging and is a risk factor for, if not a cause of, a wide spectrum of human pathophysiologic conditions.[56] In yeast, autophagy is crucial for survival during nutrient deprivation because it recycles macromolecules to provide energy and nutrients. Substrates of autophagic complexes include intracytoplasmic aggregate-prone proteins associated with neurodegenerative diseases, such as mutant α-synuclein (which contributes to PD pathogenesis),[57] neurofibrillary tangles (which contribute to AD pathogenesis),[58] and polyglutamine-expanded proteins such as mutant huntingtin (which contributes to HD pathogenesis).[51] In fact, when autophagy is impaired in mice that are otherwise healthy, protein aggregates begin to form in the cytoplasm. In addition, autophagy is critical for maintaining organellar homeostasis, especially mitochondrial health. Dysfunctional mitochondria, such as those that have lost membrane potential and are more inclined to release toxic apoptotic mediators and reactive oxygen species (ROS), are selectively removed by mitophagy compared with healthy mitochondria.[59] In addition, autophagy may also facilitate the degradation of various bacteria and viruses, and this may have a role in protecting us from a wide range of infectious diseases.[60]

The role of autophagy as a cell maintenance and turnover mechanism is particularly critical for long-lived postmitotic cells such as neurons, cardiac myocytes, and skeletal muscle fibers, which are characterized by a low replacement rate, if they are replaced at all.[61] In comparison, short-lived postmitotic cells, such as peripheral blood cells and intestinal epithelial cells, are continuously renewed through proliferation and differentiation of stem cells, allowing the dilution or easy elimination of oxidatively or otherwise damaged biological structures.[59] Damaged structures (often likened to biological garbage or waste) gradually accumulate within long-lived postmitotic cells, which suggests that they have not been properly recycled. It appears that intracellular waste accumulates both within lysosomes and generally within the cell cytoplasm,[61,62] indicative of insufficient autophagic sequestration and degradation, respectively.

Morphometric studies quantifying the number of autophagosomes are formed in different tissues of old animals, as well as their subsequent elimination, have shown a decrease in both the formation and elimination of autophagosomes with age.[63] Under some circumstances, defective autophagosome clearance may result from reduced protease activity in lysosomes with aging, or impairment of the ability of lysosomes to fuse with autophagosomes. As a result, it seems that accumulation of undigested products (in the form of autofluorescent lipofuscin aggregates made by lipids, misfolded proteins, and metals) in lysosomes may contribute to the decrease in autophagosome elimination.[64] By the end of their lives, postmitotic cells are directed primarily to abundant lipofuscin-loaded lysosomes, which leaves scant enzyme capacity for useful functions such as the autophagic turnover of organelles and macromolecules. In turn, this leads to a deterioration in the autophagic clearance of damaged structures and a decrease in functional efficiency.[65]

Damaged mitochondria and indigestible, oxidized protein aggregates are examples of extralysosomal waste. As cells age, mitochondria become enlarged and deteriorate structurally, exhibiting swelling and disintegration of cristae, which often results in the formation of amorphous material.[66] Senescent mitochondria are reported to have a diminished capacity to generate adenosine triphosphate (ATP) while producing an increased amount of ROS. Mechanisms underlying age-related mitochondrial dysfunction have been explored, and results suggest that the dysfunction may result from the deterioration of different cellular pathways, including redox homeostasis, mitochondrial and nuclear DNA repair, mitophagy, and the ubiquitin-proteasome system (UPS), among others.[67,68] It is conceivable that damaged mitochondria should be autophagocytosed and degraded, but their accumulation with age indicates an imbalance between the increase in damage attack and the decrease in recycling via mitophagy and UPS.

2. Mitochondrial dysfunction

Mitochondrial function has a central role in metabolism, apoptosis, and disease[23,25,69] and a far-reaching impact on the process of aging. As organisms and cells inevitably age over time, the efficacy of the respiratory chain tends to diminish, leading to an increase in electron leakage and a decrease in ATP generation. Considerable progress has been made in our knowledge about the relationships among mitochondrial dysfunction, energy shortages, and aging cells, which have raised interest in dissecting the details of mitochondrial dysfunction for aging research in mammals.

2.1 Mitochondrial biology

It is postulated that the mitochondrion is the result of evolution from the perfect marriage of an α-proteobacterium and a precursor of a modern eukaryotic cell, evidenced by its being the only nonnuclear subcellular organelle that has its own DNA.[70] A mitochondrion is a bilayer, semi-autonomous cellular organelle separated from the cytoplasm by mitochondrial membranes (Fig. 5.3). The outer mitochondrial membrane (OMM) is spongy, which allows for the free passage of small, uncharged metabolites and ions through voltage-dependent anion channels, or porins. Whereas larger molecules, particularly proteins, have to be transported inwardly via special translocases, the inner mitochondrial membrane (IMM) is highly impermeable and forms the main barrier between the matrix

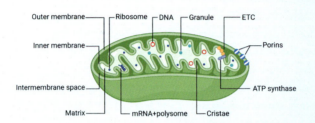

Figure 5.3 *Main structures of a mitochondrion. A mitochondrion is a bilayer, semi-autonomous cellular organelle separated from the cytoplasm by the mitochondrial membranes. It has four distinct domains: the outer mitochondrial membrane (OMM), the inner mitochondrial membrane (IMM), the intermembrane space, and the matrix. The OMM contains different types of proteins, including porins that allow the exchange of ions into and out of the mitochondrion. The IMM is differentiated into the cristae and the inner boundary membrane and contains a variety of enzymes including ATP synthase and some transport proteins. The cristae extend into the matrix and are the main sites of mitochondrial energy conversion via the electron transport chain (ETC). The space within the inner membrane is the matrix, which contains enzymes of tricarboxylic acid cycle, alongside ribosomes, DNA, RNA, polysomes, and granules.*

compartment and the cytosol. The space between the OMM and the IMM, termed the intermembrane space (IMS), harbors a variety of proteins involved in different physiologic functions (e.g., proteins for mitochondrial structural integrity and control of programmed cell death).[71] The IMM has three functional and morphologically different subregions: the cristae, the cristae junctions, and the inner boundary membranes. These structures alter their shape dynamically in response to the cell's metabolic needs and stress. The IMS and mitochondrial matrix contain proteins for the tricarboxylic acid (TCA) or Krebs cycle and the electron transport chain (ETC) that produces ATP from the redox gradient (oxidative phosphorylation [OXPHOS]), making it the major metabolic hub for maintaining cellular homeostasis and energy needs.[72] The ETC complexes encompass five protein complexes: complex I (nicotinamide adenine dinucleotide [NADH]-ubiquinone oxidoreductase), complex II (succinate–ubiquinone oxidoreductase), complex III (ubiquinol-cytochrome *c* oxidoreductase), complex IV (cytochrome *c* oxidase), and complex V (F0F1-ATP synthase), along with two mobile electron carriers, ubiquinone and cytochrome *c*.[73] Mitochondria contain their own genome with a modified genetic code, and the mammalian mitochondrial genome is passed down exclusively from the maternal germ line. Mitochondrial DNA (mtDNA) resides within the matrix and is a double-stranded, circular molecule of 16,569 base pairs containing 37 genes coding for the two ribosomal RNAs of the mitochondrial ribosome, 22 transfer RNAs required for the translation of proteins encoded by mtDNA, and 13 polypeptide subunits of the core enzyme complexes of OXPHOS.[74] Different from the nuclear genome, the mitochondrial genome encodes proteins essential for energy production, which enables mitochondria to function as the powerhouse of the cells. mtDNAs are intron-free, which makes them more susceptible to mutagenesis than nuclear DNA.[75]

2.2 Mitochondrial bioenergetics

Mitochondria power our cells by extracting energy from nutrients and distributing it to drive the machinery of life. Mitochondria have important roles in multiple basic cellular processes, such as energy production, nucleotide biosynthesis, and iron metabolism.[76] Both mitochondrial and cellular bioenergetics are affected by aging, and age-related dysfunction is implicated in many diseases, including cancer, neurodegenerative diseases, and diabetes.[77–79] Mitochondria are the main source of ATP in most cells, because many catabolic pathways converge in this organelle and result in the production of ATP by OXPHOS through the ETC. The catabolism of metabolites from glucose, protein, and fatty acids produces acetyl-CoA, which is in turn oxidized in the TCA cycle, generating the reduced electron donors NADH and flavin adenine dinucleotide.[80] Electrons are then passed through the ETC and used to create water (H_2O) by reacting hydrogen ions (H^+) with molecular oxygen (O_2). This is an exothermic reaction and releases energy used to pump protons from the outer matrix into the IMS at complexes I, III, and IV, creating an electrochemical gradient.[81] Energy released by the dissipation of this proton gradient is then coupled to the synthesis of ATP from adenosine diphosphate (ADP) and inorganic phosphate (P_i), catalyzed by ATP synthase.

2.3 Mitochondrial dysfunction in aging

Mitochondria regulate diverse biological processes for the proper survival and precise functioning of cells. They functionally control the generation of energy, the ETC, cellular signaling, and programmed cell death, whereas disorders of the mitochondria have serious health consequences and may be a root cause of multiple types of aging-associated damage.[82] Considering the involvement of mitochondria in the critical processes of cells as well as their health impacts, researchers are exploring and attempting to understand the detailed mechanisms underlying the role of mitochondrial dysfunction in aging as well as in the pathogenesis of other human diseases.

3. Mechanisms of mitochondrial dysfunction

In this section, we enumerate seven tentative mechanisms of mitochondrial dysfunction and summarize how the accumulation of dysfunctional mitochondria participates in energy shortages and aging.

3.1 Reduced mitochondrial biogenesis and integrity

Mitochondrial biogenesis is the process by which new mitochondria are formed by the growth (increase in mitochondrial mass) and division of preexisting mitochondria (increase in mitochondrial number). Mitochondrial biogenesis involves the synthesis of the IMM and OMM and mitochondrial-encoded proteins, synthesis and importation of nuclear-encoded proteins, and replication of mtDNA.[83] Mitochondrial biogenesis also requires the coordination of two genomes between the mitochondria and nucleus. Because the mitochondrial genome has limited encoding capacity, most of mitochondrial protein synthesis is under the control of the nuclear-encoded genome. A major protein controlling mitochondrial biosynthesis is peroxisome proliferator-activated receptor-γ coactivator-1α (PGC-1α).[84] After being activated by phosphorylation or deacetylation, PGC-1α activates two key nuclear transcription factors, nuclear respiratory factors 1 and 2 (NRF1 and NRF2), and subsequently mitochondrial transcription factor A (TFAM).[85] Activation of the PGC-1α−NRF−TFAM pathway leads to synthesis of mtDNA and proteins, and the generation of new mitochondria.[85]

Reduced biogenesis is strongly linked to mitochondrial dysfunction. Findings suggest that the amount and integrity of mitochondria may decline with age and lead to aberrant expression of ETC proteins, impairing

OXPHOS and energy production. With advanced age, the mitochondrial density in skeletal muscle was shown to decline gradually. This suggests a decrease in mitochondrial biogenesis.[86] The decline in mitochondrial biogenesis may result from an age-dependent reduction in levels of PGC-1α. In aged mice, overexpression of PGC-1α in skeletal muscle was associated with reduced sarcopenia (which refers to an involuntary loss of muscle mass and strength with aging) and an improvement in mitochondrial function. Impaired balance between fission and fusion events are also related to an age-dependent decline in mitochondrial biogenesis and integrity.[87] Fusion and fission of mitochondria at the organellar level are the primary molecular pathways used to control quality, which is vital because they are involved in the selection and segregation of dysfunctional mitochondria. In addition, mitophagy is used selectively to eliminate damaged mitochondria, and if mitophagy is not enough or the damage is too severe, cell death ultimately occurs.

3.2 Reduced mitochondrial dynamics

Age-related changes in mitochondria are associated with a reduction in mitochondrial function. With advanced age, mitochondrial biogenesis declines and integrity and functionality decrease. Age-dependent abnormalities in mitochondrial quality control and alterations in mitochondrial dynamics further weaken and impair mitochondrial function.[88] Mitochondrial dynamics include the mitochondrial architecture and connectivity regulated by fusion and fission events, and the locomotion of mitochondria along the cytoskeleton (mitochondrial trafficking, or motility).[89] This dynamic network is essential to ensure restructuring of the mitochondrial pool to adapt to nutrient availability, molecular signals, and cellular stress, and it is crucial to maintaining normal mitochondrial functions. Defects in mitochondrial dynamics and the signals they receive are thought to lead to their dysfunction, which is linked to aging.[90]

Fusion and fission regulate mitochondrial morphology and functionality. Mitochondrial fusion begins with the tethering of two separate mitochondria via membrane-bound dynamin-related proteins. Fusion of the outer membrane is mediated by mitofusin 1 and 2 (MFN1 and MFN2), whereas optic atrophy 1 mediates the fusion of the IMM by interacting with cardiolipin located on the opposing site of the IMM.[91] The loss of either OMM or IMM fusion proteins results in a hyperfragmented mitochondrial network characterized by the presence of many smaller mitochondria, rather than a network of a highly interconnected and elongated mitochondria. Mitochondrial fission is initiated by ER-dependent or independent mechanisms. Proteins including mitochondrial fission factor, mitochondrial fission 1 protein, mitochondrial dynamics protein (MiD)49 and MiD51, cytosolic GTPase Drp1, and dynamin-2 are recruited and orchestrated during fission.[92] Fusion and fission are processes critical for mitochondrial fitness and health. Aberrant mitochondrial structures have long been observed in aging and age-related diseases, indicating that mitochondrial dynamics are compromised as cells age. For instance, mouse knockout of genes essential for mitochondrial fusion (i.e., *Mfn1*, *Mfn2*, and *Opa1*) results in mitochondrial dysfunction and embryonic lethality.[48] Several proteins involved in mitochondrial fission are known to be dysregulated with age, which likely contributes to the altered mitochondrial network architecture seen in old organisms. In aged mice, Drp1 activity reduced and altered mitochondrial morphology in several tissues and cells, including skeletal muscle, neurons, and oocytes.[93] It was also shown that induction of Drp1p expression in *Drosophila* prolongs the life span and improves the health span via improved mitochondrial respiration and autophagy.[94] These findings indicate that restoring specific protein expression might result in improved mitochondrial function, healthier morphology, and extended organismal life span.

In addition to changes in mitochondrial shape, mitochondrial dynamics include an active process called mitochondrial trafficking.[95] Mitochondrial motility, that is to say the movement of mitochondria both intracellularly and intercellularly, is vital to various homeostatic processes. Intracellular mitochondrial motility is aided through the coordination of MFN1 and MFN2, tethering motility proteins such as MIRO and Milton, and the microtubules that make up the cytoskeletal apparatus. Mitochondrial trafficking decreases with age and contributes to mitochondrial dysfunction.[96] This decrease in mitochondrial motility is likely due to a combination of several broad changes in the cellular milieu that occur with age, but it is difficult to separate causative agents from wider symptoms associated with general decline in aging.[97] Previous studies showed that mitochondrial and cellular changes can lead to alterations in mitochondrial trafficking, including changes in mitochondrial membrane potential (MMP), and changes in cellular ATP/ADP concentrations and Ca^{2+} levels. It is expected that MMP, cellular ATP/ADP ratios, and Ca^{2+} all change as cells age, indicating that mitochondrial locomotion might be affected by a number of different processes.[98] Similarly, increased ROS production with age decreases mitochondrial trafficking through a p38α-dependent pathway.[97] Proteins such as MFN1 and MFN2 are partially interconnected, with apparently dual roles in fusion and fission, and mitochondrial movement. Loss of either *Mfn1* or *Mfn2* alters mitochondrial movement, and MFN2 expression has been found to decrease in aged muscle as well as to show a decline in mitochondrial motility.[99] In addition, both anterograde and retrograde mitochondrial trafficking decrease with age.[100] It has been well-documented that mitochondrial trafficking is altered in AD. Through axonal transport, mitochondria are transported within neurons from (anterograde) and to (retrograde) the cell body. Glycogen synthase kinase 3, a serine/threonine-protein kinase, phosphorylates serine and threonine amino acid residues on other proteins and has been shown to impair mitochondrial transport in AD by phosphorylating and deactivating mito-chondrial transport motor proteins.[101,102]

3.3 Compromised mitophagy

In cells, efficient removal of impaired and damaged mitochondria by means of a highly conserved cellular process termed mitophagy has a fundamental part in mitochondrial and metabolic homeostasis, energy supply, and the choice between cell survival and death.[23] To the contrary, defective mitophagy resulted in the accumulation of damaged mitochondria and cellular dysfunction, leading to aging and age-related neurodegeneration.

Studies on the fates and diseases of cellular organisms have greatly improved our understanding of mitochondria. It is clear that this multi-faceted organelle has crucial roles in metabolic homeostasis, regulation of Ca^{2+} and redox signaling, direct mitochondria—nucleus communications, the mediation of cell survival, stress resistance, and death.[103] Mitochondrial dysfunction is intertwined with a wide variety of human diseases. For example, certain mutations in either nuclear or mitochondria-encoded proteins can cause rare chronic and genetic mitochondrial disorders, and ATP depletion brought about by mitochondria depletion can impair cell signaling and has been associated with the etiology and pathogenesis of metabolic and neurodegenerative disease.[104] Prototypical examples include the neurodegenerative diseases AD, PD, ALS, and HD, as mentioned earlier.[23,51,105]

Cells have developed sophisticated and intertwined regulatory pathways to maintain mitochondrial homeostasis, balancing mitochondrial biogenesis with the disposal of impaired mitochondria. Mitophagy mediates the clearance of defective and superfluous mitochondria.[106] However, under certain physiologic conditions, mitophagy can be induced to degrade healthy mitochondria as well. For example, in the development of fertilized oocytes, all paternal mitochondria are eliminated, and during the maturation of erythrocytes, mitochondria (along with all other subcellular organelles) are removed from the cell.[47] Despite this, under normal conditions, mitophagy is thought to be mainly responsible for the elimination of dysfunctional mitochondria, helping to maintain basal mitochondrial turnover and acting as a quality control mechanism. In addition, mitophagy can be induced as a stress-response mechanism by various insults from the environment to remove damaged mitochondria and prevent mitochondrial-dependent apoptosis, and it can also be induced in some mitochondria-related mutants.[107]

Aging is known as a primary drivers of neurodegenerative disease; nevertheless, the underlying mechanisms remain elusive. Abnormalities and defects in mitochondria are usually concomitant with a wide range of age-related pathophysiologic features, such as impaired energy metabolism, reduced ETC function, elevated ROS levels, defective cytoplasmic calcium buffering, and increased release of proapoptotic factors. As such, mitochondrial dysfunction is considered a hallmark of aging.[108]

Over the past few decades, experimental studies using short- and long-lived model organisms, including yeasts, nematodes, flies, and rodents, have emphasized the contribution of decreased mitochondrial capacity in aging and age-associated disorders.[109] Accumulation of dysfunctional and disorganized organelles has been shown in certain tissue and cell types during the aging process. The inability of cells to clear excess or damaged organelles, likely related to age-dependent decreases in macroautophagy and mitophagy, alters the homeostatic balance of energy and exacerbates cellular dysfunction.[110]

Mitophagy is pivotal for eliminating damaged mitochondria and is necessary for maintaining a healthy mitochondrial pool. Aging, genetic, and other factors that directly or indirectly impair the mitophagy machinery lead to the accumulation of damaged mitochondria. This accumulation can lead to apoptotic activation and inflammation, and contributes to compromised mitochondria—nuclear communication. In addition to mitophagy, other known pathways are able to eliminate damaged mitochondria intracellularly (e.g., piecemeal mitophagy[111] and UPS) and extracellularly (e.g., exophers[112] and jettison protein aggregates[113]).

3.4 Accumulation of mutations in mitochondrial deoxyribonucleic acid

Age as well as genetic and environmental factors can lead to increases in mtDNA mutations, which in turn contribute to aging and aging-related disease. Although most mitochondrial proteins are encoded in and imported from the nuclear genome, mitochondria contain several copies of their own DNA.[114] Inherited mutations in mtDNA are known to cause a variety of diseases, most of which affect the brain, muscles, and other organs with a high need for energy.[115] It is well-established that mtDNA is more susceptible to mutagenesis, with a mutation rate generally 10-fold higher than that of nuclear DNA.[116] mtDNA is especially prone to large-scale deletions and point mutations, and it is considered a major target for aging-associated somatic mutations, partly owing to (1) the reduced anti-oxidative microenvironment of the mitochondria, (2) the lack of protective histones in mtDNA, and (3) the limited efficiency of mtDNA repair mechanisms.[117–119] The accumulation of deletions and point mutations with age correlates with a decline in mitochondrial function. In a model of proofreading-deficient polymerase-γ mice, the main polymerase used in mtDNA replication, mice accumulated mtDNA mutations at high levels in all tissues owing to uncorrected errors during replication.[120] This marked increase in mtDNA mutations resulted in mitochondrial dysfunction characterized by decreased respiratory enzyme activity and ATP production. In addition, these mice displayed many features common in progerias, including weight loss, kyphosis (a form of spinal deformity), anemia, gonadal atrophy and sarcopenia, and, eventually, early mortality.[120]

The role of damaged mtDNA and its detrimental effect on aging and age-related disease first came to light with the recognition that several human multisystem disorders resulted from mtDNA mutations.[121] Causative evidence for this was later provided by studies in polymerase $\gamma-$ deficient mice. These mutant mice exhibit features of premature aging and have a reduced life expectancy, and mtDNA sequencing revealed an accumulation of random point mutations and deletions in their mtDNA.[122] More generally, cells from these progeroid mice show impaired mitochondrial function. In particular, stem cells, critical for the health and regeneration of tissues, were particularly sensitive to the accumulation of mtDNA mutations, which is especially important in the context of mitochondrial dysfunction in aging. Future studies are required to confirm whether decreasing the load of mtDNA mutations via genetic manipulations is beneficial for prolonging life.

3.5 Mitochondrial oxidative phosphorylation dysfunctions

Mitochondria generate energy for the cell through OXPHOS of metabolic products to produce ATP. Under normal physiologic conditions, more than 95% of a cell's energy in the form of ATP is generated in the mitochondria by OXPHOS. As stated in Section 2.2, this process involves five different protein complexes (complexes I—V) and the movement of electrons produced by the citric acid cycle (CAC). Aside from these five complexes, more than 90 polypeptides are also involved in OXPHOS[123] and are encoded by genes from both the nuclear and mitochondrial genomes. Whereas many proteins required for the proper functioning of mitochondria are encoded by mtDNA, others are imported from the nucleus, including those involved in mtDNA replication and transcription, maintaining mtDNA and ribosomal stability, cofactors for heme, flavin, and nonheme iron, and a wide range of proteins involved in the assembly factors for each complex.[124] In all, approximately 300 proteins are required for OXPHOS to function properly. The overall process of OXPHOS is tightly controlled by transcriptional and translational regulation at the DNA and RNA levels through substrate-dependent feedback inhibition and by posttranslational modifications including phosphorylation and acetylation.[125] Mitochondrial OXPHOS is responsible for producing high-energy phosphates and is also involved in a variety of cellular processes, including intracellular calcium homeostasis, production of ROS, cAMP/protein kinase A signaling, inflammation, and apoptosis.[126] As such, it is unsurprising that various OXPHOS dysfunctions, whether due to genetic causes or lifestyle and environmental factors, have been linked to larger metabolic disorders as a whole, as well as to several late-onset human diseases and the larger aging process itself.

Aging impairs mitochondrial OXPHOS, and there is a general decrease in efficient electron transfer through complexes I—IV in the mitochondria with age, leading to a reduced capacity for aerobic energy production, while exacerbating disorders caused by impaired metabolic flexibility and promoting the generation of new such disorders. ROS levels also rise owing to malfunctions in the enzymes of the ETC, particularly complexes I, II, and III.[127]

Whereas mitochondrial dysfunction has adverse effects on aged tissues and cells, it can also have a protective role in promoting longevity, especially when targeted at earlier stages of life.[128] For instance, the reduction of nuclear-encoded OXPHOS proteins via RNA interference leads to an extended life span in *C. elegans*, but only when treatment is started in the early larval stages. Because the protein complexes of OXPHOS are encoded by both nuclear DNA and mtDNA, studies show that inhibiting mitochondrial translation and thus lowering the synthesis of mtDNA-encoded OXPHOS proteins can extend the life span in *C. elegans*. A similar effect was demonstrated in mice.[129]

It has become clear that mitochondria function as more than just an energy source for cells, and that they have a significant role in other aspects, such as organismal development, maintenance, and degeneration. In addition, OXPHOS, which is carried out in the mitochondria, is critical to maintaining cellular homeostasis. Malfunctions in OXPHOS and ATP production are linked to many diseases and dysfunctions.

3.6 High levels of reactive oxygen species

Like a double-edged sword, cellular ROS has two opposite cellular functions: whereas physiologically low doses of ROS are an important element in signal transduction, dysregulated and high levels of ROS are dangerous and can damage subcellular organelles, which can lead to a host of diseases. Here, we will focus largely on the negative effects of excess ROS, because ROS steadily increases to harmful levels as we age. Mitochondria are the primary cellular consumers of oxygen, which they use in the ETC to generate energy in the form of ATP, generating the ROS superoxide (O_2^-). Other mitochondrial enzymes are also known to generate ROS within the mitochondrial matrix, including several TCA cycle enzymes aconitase and α-ketoglutarate dehydrogenase, as well as other metabolic enzymes, such as pyruvate dehydrogenase and glycerol-3-phosphate dehydrogenase, dihydroorotate dehydrogenase, monoamine oxidases A and B, and cytochrome $b5$ reductase, along with the ETC complexes, specifically complexes I, II, and III,.[130] The generation of O_2^- depends critically on local O_2 concentrations, protonmotive force, and the $NAD^+/NADH$ and $CoQ/CoQH$ ratios.[131]

Perspectives and findings on the causal implication of ROS in aging have been controversial. The mitochondrial free radical theory of aging proposes that progressive mitochondrial dysfunction during aging leads to an increase in ROS production, which results in further mitochondrial deterioration and general cellular lesions.[96] Multiple studies support a role for ROS in aging, in which mitochondrial dysfunction can potentially be exacerbated by increasing ROS or decreasing the availability of ATP.

Mitochondria contain multiple electron carriers able to generate and produce ROS, as well as an extensive network of antioxidant defenses. Mitochondrial damage can lead to the inefficient removal of ROS, resulting in net ROS production and an imbalance of the mitochondrial ROS environment.[132] The negative impact of this net mitochondrial ROS production in aging is supported by observations that enhancing mitochondrial antioxidant defenses can increase longevity in short-lived strains of *D. melanogaster* upon overexpression of mitochondrial antioxidant enzymes such as manganese superoxide dismutase and methionine sulfoxide reductase.[133] Hence, net production of ROS is an important mechanism by which mitochondria are thought to contribute to the general physiologic decline seen in aging. However, studies have reevaluated the mitochondrial free radical theory of aging and revealed that in several models using yeast, *C. elegans*, and mice, unexpected observations showed that increasing ROS through genetic manipulation actually extends organismal longevity instead of accelerating aging.[108,134] These studies have paved the way for a fundamental reassessment of ROS in aging. There is a growing view that, in response to physiologic stress signals, intracellular ROS triggers and facilitates protective signaling cascades that cue for proliferation and survival. The most prevalent hypothesis has shaped ROS as a stress–induced molecule involved in survival signal transduction that compensates for aging-related progressive deterioration.[108] However, as chronological age advances, the rising level of ROS required to attempt to maintain survival is programmed to diminish, and ultimately exasperates rather than mitigates age-associated damage.[135] Several clinical trials focusing on treatment with antioxidants have been performed[136]; however, results have been disappointing. The reason for this may be because these antioxidant treatments have also negated the beneficial effects induced by low levels of ROS.[137]

3.7 Low levels of nicotinamide adenine dinucleotide

NAD^+ has a fundamental role in health and disease. In addition to its role as an electron carrier in metabolic pathways, NAD^+ is a cosubstrate of several proteins involved in protein posttranslational modifications, DNA repair, and several cellular bioenergetic processes.[138] These proteins are the class III histone deacetylases SIRTs, poly(ADP-ribose) polymerases (PARPs), ADP ribosyl-cyclases (CD38/CD157), and NADase sterile alpha and TIR motif-containing 1 (SARM1).[139] There are three major pathways through which NAD^+ is synthesized, including de novo biosynthesis, the

Preiss-Handler pathway, and the salvage pathway.[52] The salvage pathway is considered the most efficient method for replenishing NAD^+. NAD^+ is kept in constant equilibrium between synthesis, consumption, and recycling in many intracellular compartments, including the cytoplasm, nucleus, mitochondria, and Golgi apparatus.[51]

NAD^+ has an important role in the health span and longevity. A key role for NAD^+ in glycolysis and the TCA cycle is its ability to accept hydride equivalents, forming NADH during ATP synthesis.[140] In the mitochondria, NADH provides electrons to the ETC for OXPHOS. NAD^+/NADH ratios are critical in various bioenergetic reactions in different subcellular compartments. Changes in this ratio can have a profound impact on metabolic homeostasis. Moreover, the conversion of NAD^+ to $NADP^+$/NADPH has a vital role in a multitude of different cellular functions such as antioxidation and the generation of oxidative stress, calcium homeostasis, and the survival or death of cells.[141]

Mitochondrial function, through the ETC consumption of NADH, disrupts the NAD^+/NADH ratio, and thus it is critical to the overall bioavailability of NAD^+. Mitochondrial dysfunction can activate hypoxic signaling, such as triggering the accumulation of HIF1α under normal aerobic conditions.[142] A decline in the NAD^+/NADH ratio, also known as a pseudohypoxic state, has an important role in PGC1α/β-independent mitochondrial-nuclear communication, which contributes to the general decline in mitochondrial function with aging and disease.[143] During natural aging, lower NAD^+ levels are observed in tissues of various organisms including humans, mice, and *C. elegans*. The age-dependent reduction in NAD^+, NAD^+/NADH, and total NAD(H) levels in intact human brains from healthy volunteers was demonstrated using a noninvasive ^{31}P magnetic resonance—based NAD^+ assay.[144] In mice, there was a nearly 40% decrease in NAD^+ levels in the hippocampus of 10- to 12-month-old mice compared with 1-month-old mice.[145] Similarly, an age-dependent reduction in organismal NAD^+ in wild-type worms was shown.[24,51,146] NAD^+ depletion is observed not only during normal aging but also in accelerated aging diseases, including ataxia telangiectasia, xeroderma pigmentosum group A (XPA), Cockayne syndrome, and Werner syndrome.[24,54]

There is a close relationship between cellular (including nuclear, cytoplasmic, and mitochondrial) NAD^+ and mitochondrial homeostasis. Molecular evidence links NAD^+ depletion to mitochondrial dysfunction: (1) in the nucleus, NAD^+ depletion reduces the NAD^+/SIRT1-PGC-1α axis that regulates both mitochondrial biogenesis and mitophagy[24]; and (2) reduced mitochondrial NAD^+ could dampen mitochondrial PARP-based DNA repair via extrapolation.[147] Congruently, NAD^+ repletion can enhance cellular resilience and function, resulting in extended health span and life span in organism models.[50,148-150] For example, the NAD^+ precursor NMN could reduce the accumulation of dysfunctional mitochondria in human AD patient iPSC-derived neurons as well as in AD mouse and *C. elegans* models.[50] Furthermore, multiple NAD^+ augmentation strategies (including treatment with NAD^+ precursors NR, NMN, nicotinamide, or treatment with a PARP1 inhibitor or an SIRT1 activator) were used to improve health span and life span in accelerated aging animal models.[54,148] In addition to pharmacologic supplementation, physical exercise training can restore the age-dependent decline in NAD^+ salvage capacity in human skeletal muscle.[149,151]

Mechanisms causing defective mitochondrial bioenergetics are shown in Fig. 5.4. Mitochondrial function and aging are intricately linked, but this relationship is complicated. Regardless of whether mitochondrial dysfunction has a primary or secondary role, they interact, forming a deleterious downward spiral, mediating and amplifying aging and diseases of aging.

4. Dysfunctional mitochondria contribute to energy shortages and aging cells

4.1 Interfering with intracellular signaling

In molecular gerontology and geriatrics, mitochondria are particularly important because (1) mitochondria produce chemical energy ATP to meet cellular needs, and (2) with age, basal metabolic rate and physical performance decline as a result of the decline in mitochondrial function and activity.[152] It has long been recognized that aging in model organisms is associated with a decline in mitochondrial function, and that this decline is linked to an observed age-dependent decline in organ function. Similarly, links have been drawn between reduced mitochondrial function in humans and certain age-related diseases.[153]

It is well-established that oxidative stress is a major trigger for the accumulation of cellular damage, and that mitochondrial function has an integral role in producing ROS.[154] As such, mitochondria represent a principal target for antiaging strategies that many organisms have developed over time as they have evolved.[155] This has been shown in part by the fact that the Src homology/collagen adaptor protein, p66[shc], is a crucial

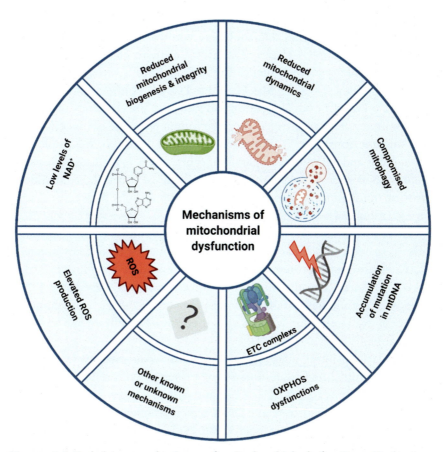

Figure 5.4 *Underlying mechanisms of mitochondrial dysfunction. Mechanisms causing defective mitochondrial bioenergetics may include (1) reduced mitochondrial biogenesis, (2) alterations in mitochondrial dynamics resulting from imbalance of fission and fusion events, (3) defective quality control by mitophagy, (4) accumulation of mutations and damage in mitochondrial DNA (mtDNA), (5) decreased oxidation of mitochondrial proteins (OXPHOS), (6) high levels of reactive oxygen species (ROS), and (7) low levels of oxidized nicotinamide adenine dinucleotide (NAD^{+}), among other known or unknown mechanisms. ETC, electron transport chain.*

regulator of mitochondrial function and acts as a master regulator of ROS production and oxidative stress in mammals. A significantly longer life expectancy was achieved by knockdown of this gene.[156] In addition, caloric restriction (CR) has been shown to ameliorate mitochondrial oxidative stress by modulating the IGF-1, mTOR, SIRTs, and AMPK signaling pathways.

Many factors contribute to the life span, all of which are part of a complex system of signals that modulate common targets in coordinated ways.[157] For example, autophagy, a critical element of the CR-mediated extension of life span, is influenced by signals from a variety of signaling molecules, such as FOXOs, SIRTs, and mTOR.[158] Other signaling inputs converge on mitochondrial function and influence the efficiency of positive and negative cellular challenges. A primary input is the rise of mitochondrial free calcium ions (Ca^{2+}). Research has shed light on the molecular mechanisms and pathophysiologic significance of Ca^{2+}-mediated effects on apoptosis and necrosis by inducing the opening of permeability transition pores of mitochondria, which have long-standing links between calcium dysregulation and cell death.[159]

There is considerable interest in the mitochondrial–nuclear signaling pathway as well, which has also been shown to be important. For instance, telomeres are nucleoprotein complexes at the end of chromosomes that work to maintain chromosomal integrity and curtail p53-mediated DNA damage. Dysfunctional telomeres can adversely affect mitochondrial biogenesis by interfering with p53 or the PGC-1 family of transcription coactivators.[160]

Cells lacking intact nuclear excision DNA repair capacity (e.g., XPA) also exhibit mitochondrial dysfunction as a result of impaired mitophagy, as evidenced by decreased NAD^+ levels and SIRT activity.[161] XPA and the related Cockayne syndrome are DNA repair disorders that lead to accelerated aging,[24,162] and there is growing evidence that diseases associated with primary nuclear DNA repair affect metabolism profoundly. Signaling occurs between the nucleus and mitochondria in both directions. Further dissection of these pathways will likely provide valuable insight into organismal aging.

4.2 Triggering apoptosis and inflammatory reactions

The aging process as well as the pathogenesis of several diseases are commonly attributed to increased ROS generation, disturbances in energy metabolism, and the crucial role of mitochondria in both necrotic and apoptotic cell death.[163] Studies have shown that mitochondrial ROS production has an important role in regulating inflammatory responses. Mitochondrial dysfunction increases the production of ROS, leading to activation of the nuclear factor-κB (NF-κB) pathway. NF-κB signaling induces the expression of proinflammatory cytokines interleukin (IL)-1β and IL-18, which are required for priming nucleotide-binding oligomerization domain–like receptor protein 3 (NLRP3), the major immune sensor for cellular stress signals, mediated by inflammasomes.[164]

A common observation in mammalian aging studies is decreased electron transfer in the mitochondria, along with a decreased capacity for energy supply, resulting in disturbances in energy metabolism in older animals. In rodent aging models, mitochondria isolated from organs and tissues with high metabolic rates of old and senescent rats and mice, including brain, heart, liver, and kidney, showed decreased electron transfer activities in complexes I and IV and an increased rate of O_2^- generation. This observed age-dependent decrease in electron transfer in mitochondria is concomitant with the development of a mitochondrial subpopulation with increased size and fragility, and decreased membrane potential.[165]

In humans, an increase in cell death mediated by age may result in a dramatic decrease in cell numbers. Mitochondria have a crucial role in executing cell death pathways.[166] Depending on the range of the insult and the amount of cytosolic Ca^{2+} (released by some organelles acting as intracellular calcium ion reservoirs, such as the Golgi apparatus and ER), either necrosis or apoptosis is activated.[167] On the one hand, cellular Ca^{2+} overload represents a threat to cellular life, because many proteases and phospholipases are activated by Ca^{2+}, so an overload of this ion can lead to necrotic cell death. On the other hand, apoptotic cell death relies on increased Ca^{2+} concentrations mediated both by ER calcium release and active calcium recruitment via mitochondria-associated membranes, in which where the two organelles organize dynamic contacts.[168,169] Bcl-2 family proteins, located at the ER and mitochondrial surfaces, and caspase family proteins are important regulators of the apoptotic process as well.[170] It has been suggested that mitochondrial dysfunction and oxidative stress could induce mitochondrial permeability, cytochrome c release, and caspase activation, leading to the initiation of mitochondria-driven apoptosis.[122] From brain studies of young and aged mice, in certain brain regions such as the hippocampus, neurons and astrocytes showed an accumulation of dysfunctional mitochondria during aging,[171] which were linked to increased apoptosis and tissue atrophy.[172]

Together with mitochondrial dysfunction, chronic inflammation is a hallmark of both the aging process and larger age-related diseases.[67] Inflammation is a conserved protective mechanism that protects organisms from acute and local perturbations and provides an adaptive response when the organisms experience injury or infection.[173] Chronic, systemic inflammation is often seen in aging, contributing generally to other aging hallmarks to such an extent that the phenomenon has been termed inflammaging. In some instances, damage to mitochondria and their general dysfunction may lead to the secretion of mtDNA and its subsequent recognition by immune pathogen recognition receptors (PRRs) as damage-associated molecular patterns (DAMPs). This may function as an important link between mitochondrial dysfunction and inflammaging.[174] mtDNA, which is released as a result of cellular and/or metabolic stress, contains hypomethylated CpG motifs resembling those of bacterial DNA, and thus is able to induce different inflammatory responses through three different PRR signaling pathways by binding and/or activating (1) TLR9 located in cell endosomes,[175] (2) the NLRP3 inflammasome,[176] and (3) the cytosolic cyclic guanosine monophosphate (GMP)-AMP synthase-stimulator of interferon genes DNA sensing system.[177] The underlying mechanisms of the contribution of mitochondrial DAMP to the inflammatory milieu that characterizes aging are incompletely understood. However, cytokines, chemokines, nitric oxide, and ROS released into the circulatory system by inflammatory cells, as well as mitochondrial DAMPs, if not properly controlled, can induce further mitochondrial and tissue damage, promoting a positive feedback loop.[178]

5. Conclusions and future perspectives

Numerous studies using laboratory animals, human tissues, and clinical trials support the following conclusions: (1) autophagy and mitophagy decline with age, (2) autophagy and mitophagy are critical processes for cellular health, and (3) impairment of autophagy or mitophagy contributes to pathologic aging and disease. Considering the wide spectrum of pathways and mechanisms associated with these processes and their manifold downstream effects, compromised autophagy is undoubtedly a hallmark of aging. Although autophagy and aging are often described as having an antagonistic relationship in which increased autophagy is beneficial and decreased autophagy is harmful, this may be overly simplistic. The key to long-term health will likely be the ability to attain the right balance of autophagy, which could be determined by the cell type, microenvironment, exercise, diet, and other known and unknown parameters. The degree and balance of both autophagy and mitophagy in each tissue vary greatly with time and the stage of life. Therefore, understanding how these dynamic changes occur over time will be crucial for healthy aging.

Acknowledgments

The authors acknowledge the valuable work of the many investigators whose published articles they were unable to cite owing to space limitations. We thank Fang Lab member Dawn Patrick-Brown for reading the paper. Figures were generated using the subscribed software BioRender, © biorender.com, Toronto, Ontario, Canada.

Funding

E.F.F. was supported by HELSE SØR-ØST (Nos. 2017056, 2020001, and 2021021), the Research Council of Norway (No. 262175), the National Natural Science Foundation of China (No. 81971327), Akershus University Hospital (Nos. 269901, 261973, and 262960), the Civitan Norges Forskningsfond for Alzheimer's sykdom (No. 281931), the Czech Republic—Norway KAPPA program (with Martin Vyhnálek, No. TO01000215), and the Rosa sløyfe/Norwegian Cancer Society and Norwegian Breast Cancer Society (No. 207819). H.L.W. was sponsored by the China Scholarship Council [www.csc.edu.cn/].

Conflict of interests

E.F.F. has a CRADA arrangement with ChromaDex (United States) and is a consultant for Aladdin Healthcare Technologies (United Kingdom and Germany), the Vancouver Dementia Prevention Center (Canada), Intellectual Labs (Norway), and MindRank AI (China).

References

1. Leidal AM, Levine B, Debnath J. Autophagy and the cell biology of age-related disease. *Nat Cell Biol.* 2018;20:1338—1348.
2. Partridge L, Deelen J, Slagboom PE. Facing up to the global challenges of ageing. *Nature.* 2018;561:45—56.
3. Fang EF, Xie C, Schenkel JA, et al. A research agenda for ageing in China in the 21st century (2nd edition): focusing on basic and translational research, long-term care, policy and social networks. *Ageing Res Rev.* 2020;64:101174.
4. Dikic I, Elazar Z. Mechanism and medical implications of mammalian autophagy. *Nat Rev Mol Cell Biol.* 2018;19:349—364.
5. Rubinsztein DC, Marino G, Kroemer G. Autophagy and aging. *Cell.* 2011;146:682—695.
6. Ashford TP, Porter KR. Cytoplasmic components in hepatic cell lysosomes. *J Cell Biol.* 1962;12:198—202.
7. Klionsky DJ. Autophagy revisited: a conversation with Christian de Duve. *Autophagy.* 2008;4:740—743.
8. Tsukada M, Ohsumi Y. Isolation and characterization of autophagy-defective mutants of *Saccharomyces cerevisiae*. *FEBS Lett.* 1993;333:169—174.
9. Kaushik S, Cuervo AM. The coming of age of chaperone-mediated autophagy. *Nat Rev Mol Cell Biol.* 2018;19:365—381.
10. Aman Y, Schmauck-Medina T, Hansen M, et al. Autophagy in healthy aging and disease. *Nat Aging.* 2021;1:634—650.
11. Pankiv S, Clausen TH, Lamark T, et al. p62/SQSTM1 binds directly to Atg8/LC3 to facilitate degradation of ubiquitinated protein aggregates by autophagy. *J Biol Chem.* 2007;282:24131—24145.
12. Zaffagnini G, Martens S. Mechanisms of selective autophagy. *J Mol Biol.* 2016;428:1714—1724.

13. Thumm M, Egner R, Schlumpberger M, Straub M, Veenhuis M, Wolf DH. Isolation of autophagocytosis mutants of *Saccharomyces cerevisiae*. *FEBS Lett.* 1994;349:275—280.
14. Hansen M, Rubinsztein DC, Walker DW. Autophagy as a promoter of longevity: insights from model organisms. *Nat Rev Mol Cell Biol.* 2018;19:579—593.
15. Bjørkøy G, Lamark T, Brech A, et al. p62/SQSTM1 forms protein aggregates degraded by autophagy and has a protective effect on huntingtin-induced cell death. *J Cell Biol.* 2005;171:603—614.
16. Kirkin V. History of the selective autophagy research: how did it begin and where does it stand today? *J Mol Biol.* 2020;432:3—27.
17. Deter RL, De Duve C. Influence of glucagon, an inducer of cellular autophagy, on some physical properties of rat liver lysosomes. *J Cell Biol.* 1967;33:437—449.
18. Gatica D, Lahiri V, Klionsky DJ. Cargo recognition and degradation by selective autophagy. *Nat Cell Biol.* 2018;20:233—242.
19. El-Houjeiri L, Possik E, Vijayaraghavan T, et al. The transcription factors TFEB and TFE3 link the FLCN-AMPK signaling Axis to innate immune response and pathogen resistance. *Cell Rep.* 2019;26:3613—3628 e3616.
20. Settembre C, Zoncu R, Medina DL, et al. A lysosome-to-nucleus signalling mechanism senses and regulates the lysosome via mTOR and TFEB. *EMBO J.* 2012;31:1095—1108.
21. Medina DL, Di Paola S, Peluso I, et al. Lysosomal calcium signalling regulates autophagy through calcineurin and TFEB. *Nat Cell Biol.* 2015;17:288—299.
22. Johansen T, Lamark T. Selective autophagy: ATG8 family proteins, LIR motifs and cargo receptors. *J Mol Biol.* 2020;432:80—103.
23. Lou G, Palikaras K, Lautrup S, Scheibye-Knudsen M, Tavernarakis N, Fang EF. Mitophagy and neuroprotection. *Trends Mol Med.* 2020;26:8—20.
24. Fang EF, Scheibye-Knudsen M, Brace LE, et al. Defective mitophagy in XPA via PARP-1 hyperactivation and NAD(+)/SIRT1 reduction. *Cell.* 2014;157:882—896.
25. Menzies FM, Fleming A, Rubinsztein DC. Compromised autophagy and neurodegenerative diseases. *Nat Rev Neurosci.* 2015;16:345—357.
26. Simonsen A, Cumming RC, Brech A, Isakson P, Schubert DR, Finley KD. Promoting basal levels of autophagy in the nervous system enhances longevity and oxidant resistance in adult Drosophila. *Autophagy.* 2008;4:176—184.
27. Sun Y, Li M, Zhao D, Li X, Yang C, Wang X. Lysosome activity is modulated by multiple longevity pathways and is important for lifespan extension in *C. elegans*. *Elife.* 2020;9.
28. Pyo JO, Yoo SM, Ahn HH, et al. Overexpression of Atg5 in mice activates autophagy and extends lifespan. *Nat Commun.* 2013;4:2300.
29. Gelino S, Chang JT, Kumsta C, et al. Intestinal autophagy improves healthspan and longevity in *C. elegans* during dietary restriction. *PLoS Genet.* 2016;12:e1006135.
30. Chang JT, Kumsta C, Hellman AB, Adams LM, Hansen M. Spatiotemporal regulation of autophagy during *Caenorhabditis elegans* aging. *Elife.* 2017;6.
31. Minnerly J, Zhang J, Parker T, Kaul T, Jia K. The cell non-autonomous function of ATG-18 is essential for neuroendocrine regulation of *Caenorhabditis elegans* lifespan. *PLoS Genet.* 2017;13:e1006764.
32. Bai H, Kang P, Hernandez AM, Tatar M. Activin signaling targeted by insulin/dFOXO regulates aging and muscle proteostasis in Drosophila. *PLoS Genet.* 2013;9:e1003941.
33. Carnio S, LoVersco F, Baraibar MA, et al. Autophagy impairment in muscle induces neuromuscular junction degeneration and precocious aging. *Cell Rep.* 2014;8:1509—1521.
34. Horvath S. DNA methylation age of human tissues and cell types. *Genome Biol.* 2013;14:R115.

Autophagy and bioenergetics in aging **139**

35. Sarkis GJ, Ashcom JD, Hawdon JM, Jacobson LA. Decline in protease activities with age in the nematode *Caenorhabditis elegans*. *Mech Ageing Dev*. 1988;45:191−201.
36. Hughes AL, Gottschling DE. An early age increase in vacuolar pH limits mitochondrial function and lifespan in yeast. *Nature*. 2012;492:261−265.
37. Wilhelm T, Byrne J, Medina R, et al. Neuronal inhibition of the autophagy nucleation complex extends life span in post-reproductive *C. elegans*. *Genes Dev*. 2017;31:1561−1572.
38. Maruzs T, Simon-Vecsei Z, Kiss V, Csizmadia T, Juhasz G. On the fly: recent progress on autophagy and aging in Drosophila. *Front Cell Dev Biol*. 2019;7:140.
39. Kaushik S, Arias E, Kwon H, et al. Loss of autophagy in hypothalamic POMC neurons impairs lipolysis. *EMBO Rep*. 2012;13:258−265.
40. Ott C, Konig J, Hohn A, Jung T, Grune T. Macroautophagy is impaired in old murine brain tissue as well as in senescent human fibroblasts. *Redox Biol*. 2016;10:266−273.
41. Triplett JC, Tramutola A, Swomley A, et al. Age-related changes in the proteostasis network in the brain of the naked mole-rat: implications promoting healthy longevity. *Biochim Biophys Acta*. 2015;1852:2213−2224.
42. Yu Y, Feng L, Li J, et al. The alteration of autophagy and apoptosis in the hippocampus of rats with natural aging-dependent cognitive deficits. *Behav Brain Res*. 2017;334:155−162.
43. Pareja-Cajiao M, Gransee HM, Stowe JM, Rana S, Sieck GC, Mantilla CB. Age-related impairment of autophagy in cervical motor neurons. *Exp Gerontol*. 2021;144:111193.
44. Fernando R, Castro JP, Flore T, Deubel S, Grune T, Ott C. Age-related maintenance of the autophagy-lysosomal system is dependent on skeletal muscle type. *Oxid Med Cell Longev*. 2020;2020:4908162.
45. Lipinski MM, Zheng B, Lu T, et al. Genome-wide analysis reveals mechanisms modulating autophagy in normal brain aging and in Alzheimer's disease. *Proc Natl Acad Sci U S A*. 2010;107:14164−14169.
46. Lionaki E, Markaki M, Palikaras K, Tavernarakis N. Mitochondria, autophagy and age-associated neurodegenerative diseases: new insights into a complex interplay. *Biochim Biophys Acta*. 2015;1847:1412−1423.
47. Onishi M, Yamano K, Sato M, Matsuda N, Okamoto K. Molecular mechanisms and physiological functions of mitophagy. *EMBO J*. 2021;40:e104705.
48. Shirihai OS, Song M, Dorn GW. 2nd How mitochondrial dynamism orchestrates mitophagy. *Circ Res*. 2015;116:1835−1849.
49. Kerr JS, Adriaanse BA, Greig NH, et al. Mitophagy and Alzheimer's disease: cellular and molecular mechanisms. *Trends Neurosci*. 2017;40:151−166.
50. Fang EF, Hou Y, Palikaras K, et al. Mitophagy inhibits amyloid-beta and tau pathology and reverses cognitive deficits in models of Alzheimer's disease. *Nat Neurosci*. 2019;22:401−412.
51. Lautrup S, Sinclair DA, Mattson MP, Fang EF. NAD(+) in brain aging and neurodegenerative disorders. *Cell Metab*. 2019;30:630−655.
52. Rana A, Rera M, Walker DW. Parkin overexpression during aging reduces proteotoxicity, alters mitochondrial dynamics, and extends lifespan. *Proc Natl Acad Sci U S A*. 2013;110:8638−8643.
53. Rana A, Oliveira MP, Khamoui AV, et al. Promoting Drp1-mediated mitochondrial fission in midlife prolongs healthy lifespan of *Drosophila melanogaster*. *Nat Commun*. 2017;8:448.
54. Fang EF, Hou Y, Lautrup S, et al. NAD(+) augmentation restores mitophagy and limits accelerated aging in Werner syndrome. *Nat Commun*. 2019;10:5284.

55. Ryu D, Mouchiroud L, Andreux PA, et al. Urolithin A induces mitophagy and prolongs lifespan in *C. elegans* and increases muscle function in rodents. *Nat Med.* 2016;22:879–888.

56. Reiten OK, Wilvang MA, Mitchell SJ, Hu Z, Fang EF. Preclinical and clinical evidence of NAD(+) precursors in health, disease, and ageing. *Mech Ageing Dev.* 2021;199:111567.

57. Cuervo AM, Bergamini E, Brunk UT, Dröge W, Ffrench M, Terman A. Autophagy and aging: the importance of maintaining "clean" cells. *Autophagy.* 2005;1:131–140.

58. Thal DR, Del Tredici K, Braak H. Neurodegeneration in normal brain aging and disease. *Sci Aging Knowledge Environ.* 2004:pe26, 2004.

59. Sheldrake AR. The ageing, growth and death of cells. *Nature.* 1974;250:381–385.

60. Levine B, Deretic V. Unveiling the roles of autophagy in innate and adaptive immunity. *Nat Rev Immunol.* 2007;7:767–777.

61. Terman A. Garbage catastrophe theory of aging: imperfect removal of oxidative damage? *Redox Rep.* 2001;6:15–26.

62. Terman A, Dalen H, Eaton JW, Neuzil J, Brunk UT. Mitochondrial recycling and aging of cardiac myocytes: the role of autophagocytosis. *Exp Gerontol.* 2003;38:863–876.

63. Stupina AS, Terman AK, Kvitnitskaia-Ryzhova T, Mezhiborskaia NA, Zherebitskii VA. [The age-related characteristics of autophagocytosis in different tissues of laboratory animals]. *Tsitol Genet.* 1994;28:15–20.

64. Terman A, Brunk UT. Lipofuscin. *Int J Biochem Cell Biol.* 2004;36:1400–1404.

65. Brunk UT, Terman A. The mitochondrial-lysosomal axis theory of aging: accumulation of damaged mitochondria as a result of imperfect autophagocytosis. *Eur J Biochem.* 2002;269:1996–2002.

66. Coleman R, Silbermann M, Gershon D, Reznick AZ. Giant mitochondria in the myocardium of aging and endurance-trained mice. *Gerontology.* 1987;33:34–39.

67. Sun N, Youle RJ, Finkel T. The mitochondrial basis of aging. *Mol Cell.* 2016;61:654–666.

68. Scheibye-Knudsen M, Fang EF, Croteau DL, Wilson 3rd DM, Bohr VA. Protecting the mitochondrial powerhouse. *Trends Cell Biol.* 2015;25:158–170.

69. Pickrell AM, Youle RJ. The roles of PINK1, parkin, and mitochondrial fidelity in Parkinson's disease. *Neuron.* 2015;85:257–273.

70. Boore JL. Animal mitochondrial genomes. *Nucleic Acids Res.* 1999;27:1767–1780.

71. Sahin E, DePinho RA. Axis of ageing: telomeres, p53 and mitochondria. *Nat Rev Mol Cell Biol.* 2012;13:397–404.

72. Kroemer G, Galluzzi L, Brenner C. Mitochondrial membrane permeabilization in cell death. *Physiol Rev.* 2007;87:99–163.

73. Green DR, Galluzzi L, Kroemer G. Mitochondria and the autophagy-inflammation-cell death axis in organismal aging. *Science.* 2011;333:1109–1112.

74. Raffaello A, Rizzuto R. Mitochondrial longevity pathways. *Biochim Biophys Acta.* 2011;1813:260–268.

75. Bratic A, Larsson NG. The role of mitochondria in aging. *J Clin Invest.* 2013;123:951–957.

76. Gershoni M, Templeton AR, Mishmar D. Mitochondrial bioenergetics as a major motive force of speciation. *Bioessays.* 2009;31:642–650.

77. Boland ML, Chourasia AH, Macleod KF. Mitochondrial dysfunction in cancer. *Front Oncol.* 2013;3:292.

78. Johri A, Beal MF. Mitochondrial dysfunction in neurodegenerative diseases. *J Pharmacol Exp Ther.* 2012;342:619–630.

79. Sivitz WI, Yorek MA. Mitochondrial dysfunction in diabetes: from molecular mechanisms to functional significance and therapeutic opportunities. *Antioxid Redox Signal.* 2010;12:537—577.
80. Martinez-Reyes I, Chandel NS. Mitochondrial TCA cycle metabolites control physiology and disease. *Nat Commun.* 2020;11:102.
81. Kuhlbrandt W. Structure and function of mitochondrial membrane protein complexes. *BMC Biol.* 2015;13:89.
82. Sorrentino V, Menzies KJ, Auwerx J. Repairing mitochondrial dysfunction in disease. *Annu Rev Pharmacol Toxicol.* 2018;58:353—389.
83. Tsiloulis T, Watt MJ. Exercise and the regulation of adipose tissue metabolism. *Prog Mol Biol Transl Sci.* 2015;135:175—201.
84. LeBleu VS, O'Connell JT, Gonzalez Herrera KN, et al. PGC-1alpha mediates mitochondrial biogenesis and oxidative phosphorylation in cancer cells to promote metastasis. *Nat Cell Biol.* 2014;16:992—1003, 1001-1015.
85. Kang D, Kim SH, Hamasaki N. Mitochondrial transcription factor A (TFAM): roles in maintenance of mtDNA and cellular functions. *Mitochondrion.* 2007;7:39—44.
86. Groennebaek T, Vissing K. Impact of resistance training on skeletal muscle mitochondrial biogenesis, content, and function. *Front Physiol.* 2017;8:713.
87. Liu YJ, McIntyre RL, Janssens GE, Houtkooper RH. Mitochondrial fission and fusion: a dynamic role in aging and potential target for age-related disease. *Mech Ageing Dev.* 2020;186:111212.
88. Ni HM, Williams JA, Ding WX. Mitochondrial dynamics and mitochondrial quality control. *Redox Biol.* 2015;4:6—13.
89. Tilokani L, Nagashima S, Paupe V, Prudent J. Mitochondrial dynamics: overview of molecular mechanisms. *Essays Biochem.* 2018;62:341—360.
90. Di Nottia M, Verrigni D, Torraco A, Rizza T, Bertini E, Carrozzo R. Mitochondrial dynamics: molecular mechanisms, related primary mitochondrial disorders and therapeutic approaches. *Genes.* 2021;12.
91. Ishihara N, Eura Y, Mihara K. Mitofusin 1 and 2 play distinct roles in mitochondrial fusion reactions via GTPase activity. *J Cell Sci.* 2004;117:6535—6546.
92. Westermann B. Mitochondrial fusion and fission in cell life and death. *Nat Rev Mol Cell Biol.* 2010;11:872—884.
93. Udagawa O, Ishihara T, Maeda M, et al. Mitochondrial fission factor Drp1 maintains oocyte quality via dynamic rearrangement of multiple organelles. *Curr Biol.* 2014;24:2451—2458.
94. Mulakkal NC, Nagy P, Takats S, Tusco R, Juhász G, Nezis IP. Autophagy in Drosophila: from historical studies to current knowledge. *Biomed Res Int.* 2014:273473, 2014.
95. Bereiter-Hahn J. Mitochondrial dynamics in aging and disease. *Prog Mol Biol Transl Sci.* 2014;127:93—131.
96. Cui H, Kong Y, Zhang H. Oxidative stress, mitochondrial dysfunction, and aging. *J Signal Transduct.* 2012;2012:646354.
97. Sharma A, Smith HJ, Yao P, Mair WB. Causal roles of mitochondrial dynamics in longevity and healthy aging. *EMBO Rep.* 2019;20:e48395.
98. Chang DTW, Reynolds IJ. Differences in mitochondrial movement and morphology in young and mature primary cortical neurons in culture. *Neuroscience.* 2006;141:727—736.
99. Sebastián D, Sorianello E, Segalés J, et al. Mfn2 deficiency links age-related sarcopenia and impaired autophagy to activation of an adaptive mitophagy pathway. *EMBO J.* 2016;35:1677—1693.

100. Milde S, Adalbert R, Elaman MH, Coleman MP. Axonal transport declines with age in two distinct phases separated by a period of relative stability. *Neurobiol Aging.* 2015;36:971–981.
101. Flannery PJ, Trushina E. Mitochondrial dynamics and transport in Alzheimer's disease. *Mol Cell Neurosci.* 2019;98:109–120.
102. Kobro-Flatmoen A, Lagartos-Donate MJ, Aman Y, Edison P, Witter MP, Fang EF. Re-emphasizing early Alzheimer's disease pathology starting in select entorhinal neurons, with a special focus on mitophagy. *Ageing Res Rev.* 2021;67:101307.
103. Fulda S, Gorman AM, Hori O, Samali A. Cellular stress responses: cell survival and cell death. *Int J Cell Biol.* 2010;2010:214074.
104. Martinez-Vicente M. Neuronal mitophagy in neurodegenerative diseases. *Front Mol Neurosci.* 2017;10:64.
105. Lautrup S, Lou G, Aman Y, Nilsen H, Tao J, Fang EF. Microglial mitophagy mitigates neuroinflammation in Alzheimer's disease. *Neurochem Int.* 2019;129:104469.
106. Ashrafi G, Schwarz TL. The pathways of mitophagy for quality control and clearance of mitochondria. *Cell Death Differ.* 2013;20:31–42.
107. Gustafsson AB, Dorn GW. 2nd evolving and expanding the roles of mitophagy as a homeostatic and pathogenic process. *Physiol Rev.* 2019;99:853–892.
108. Lopez-Otin C, Blasco MA, Partridge L, Serrano M, Kroemer G. The hallmarks of aging. *Cell.* 2013;153:1194–1217.
109. Palikaras K, Lionaki E, Tavernarakis N. Mechanisms of mitophagy in cellular homeostasis, physiology and pathology. *Nat Cell Biol.* 2018;20:1013–1022.
110. Palikaras K, Daskalaki I, Markaki M, Tavernarakis N. Mitophagy and age-related pathologies: development of new therapeutics by targeting mitochondrial turnover. *Pharmacol Ther.* 2017;178:157–174.
111. Le Guerroué F, Eck F, Jung J, et al. Autophagosomal content profiling reveals an LC3C-dependent piecemeal mitophagy pathway. *Mol Cell.* 2017;68:786–796 e786.
112. Nicolás-Ávila JA, Lechuga-Vieco AV, Esteban-Martínez L, et al. A network of macrophages supports mitochondrial homeostasis in the heart. *Cell.* 2020;183:94–109 e123.
113. Melentijevic I, Toth ML, Arnold ML, et al. *C. elegans* neurons jettison protein aggregates and mitochondria under neurotoxic stress. *Nature.* 2017;542:367–371.
114. Taanman JW. The mitochondrial genome: structure, transcription, translation and replication. *Biochim Biophys Acta.* 1999;1410:103–123.
115. Magistretti PJ, Allaman I. A cellular perspective on brain energy metabolism and functional imaging. *Neuron.* 2015;86:883–901.
116. Haag-Liautard C, Coffey N, Houle D, Lynch M, Charlesworth B, Keightley PD. Direct estimation of the mitochondrial DNA mutation rate in *Drosophila melanogaster*. *PLoS Biol.* 2008;6:e204.
117. Alexeyev M, Shokolenko I, Wilson G, LeDoux S. The maintenance of mitochondrial DNA integrity–critical analysis and update. *Cold Spring Harb Perspect Biol.* 2013;5:a012641.
118. Mailloux RJ, Jin X, Willmore WG. Redox regulation of mitochondrial function with emphasis on cysteine oxidation reactions. *Redox Biol.* 2014;2:123–139.
119. Singh G, Pachouri UC, Khaidem DC, Kundu A, Chopra C, Singh P. Mitochondrial DNA damage and diseases. *F1000Res.* 2015;4:176.
120. Trifunovic A. Mitochondrial DNA and ageing. *Biochim Biophys Acta.* 2006;1757:611–617.
121. Zapico SC, Ubelaker DH. mtDNA mutations and their role in aging, diseases and forensic sciences. *Aging Dis.* 2013;4:364–380.
122. Lin MT, Beal MF. Mitochondrial dysfunction and oxidative stress in neurodegenerative diseases. *Nature.* 2006;443:787–795.

123. Mai N, Chrzanowska-Lightowlers ZM, Lightowlers RN. The process of mammalian mitochondrial protein synthesis. *Cell Tissue Res.* 2017;367:5—20.
124. Fernandez-Vizarra E, Zeviani M. Mitochondrial disorders of the OXPHOS system. *FEBS Lett.* 2021;595:1062—1106.
125. Duarte FV, Palmeira CM, Rolo AP. The role of microRNAs in mitochondria: small players acting wide. *Genes.* 2014;5:865—886.
126. Bergman O, Ben-Shachar D. Mitochondrial oxidative phosphorylation system (OXPHOS) deficits in schizophrenia: possible interactions with cellular processes. *Can J Psychiat.* 2016;61:457—469.
127. Bhatti JS, Bhatti GK, Reddy PH. Mitochondrial dysfunction and oxidative stress in metabolic disorders - a step towards mitochondria based therapeutic strategies. *Biochim Biophys Acta Mol Basis Dis.* 2017;1863:1066—1077.
128. Srivastava S. The mitochondrial basis of aging and age-related disorders. *Genes.* 2017;8.
129. Houtkooper RH, Mouchiroud L, Ryu D, et al. Mitonuclear protein imbalance as a conserved longevity mechanism. *Nature.* 2013;497:451—457.
130. Starkov AA. The role of mitochondria in reactive oxygen species metabolism and signaling. *Ann N Y Acad Sci.* 2008;1147:37—52.
131. Murphy MP. How mitochondria produce reactive oxygen species. *Biochem J.* 2009;417:1—13.
132. Stowe DF, Camara AK. Mitochondrial reactive oxygen species production in excitable cells: modulators of mitochondrial and cell function. *Antioxid Redox Signal.* 2009;11:1373—1414.
133. Bayne AC, Mockett RJ, Orr WC, Sohal RS. Enhanced catabolism of mitochondrial superoxide/hydrogen peroxide and aging in transgenic Drosophila. *Biochem J.* 2005;391:277—284.
134. Pomatto LCD, Davies KJA. Adaptive homeostasis and the free radical theory of ageing. *Free Radic Biol Med.* 2018;124:420—430.
135. Hekimi S, Lapointe J, Wen Y. Taking a "good" look at free radicals in the aging process. *Trends Cell Biol.* 2011;21:569—576.
136. Forman HJ, Zhang H. Targeting oxidative stress in disease: promise and limitations of antioxidant therapy. *Nat Rev Drug Discov.* 2021;20:689—709.
137. Ristow M. Unraveling the truth about antioxidants: mitohormesis explains ROS-induced health benefits. *Nat Med.* 2014;20:709—711.
138. Xie N, Lu Z, Huang C, et al. NAD(+) metabolism: pathophysiologic mechanisms and therapeutic potential. *Signal Transduct Target Ther.* 2020;5:227.
139. Fang EF, Lautrup S, Hou Y, et al. NAD(+) in aging: molecular mechanisms and translational implications. *Trends Mol Med.* 2017;23:899—916.
140. Canto C, Menzies KJ, Auwerx J. NAD(+) metabolism and the control of energy homeostasis: a balancing act between mitochondria and the nucleus. *Cell Metab.* 2015;22:31—53.
141. Ying W. NAD+/NADH and NADP+/NADPH in cellular functions and cell death: regulation and biological consequences. *Antioxid Redox Signal.* 2008;10:179—206.
142. Lee CF, Caudal A, Abell L, Nagana Gowda GA, Tian R. Targeting NAD(+) metabolism as interventions for mitochondrial disease. *Sci Rep.* 2019;9:3073.
143. Gomes AP, Price NL, Ling AJY, et al. Declining NAD(+) induces a pseudohypoxic state disrupting nuclear-mitochondrial communication during aging. *Cell.* 2013;155:1624—1638.
144. Zhu XH, Lu M, Lee BY, Ugurbil K, Chen W. In vivo NAD assay reveals the intracellular NAD contents and redox state in healthy human brain and their age dependences. *Proc Natl Acad Sci U S A.* 2015;112:2876—2881.
145. Stein LR, Imai S. Specific ablation of Nampt in adult neural stem cells recapitulates their functional defects during aging. *EMBO J.* 2014;33:1321—1340.

146. Mouchiroud L, Houtkooper RH, Moullan N, et al. The NAD(+)/Sirtuin pathway modulates longevity through activation of mitochondrial UPR and FOXO signaling. *Cell.* 2013;154:430–441.
147. de Graaf RA, Behar KL. Detection of cerebral NAD(+) by in vivo (1)H NMR spectroscopy. *NMR Biomed.* 2014;27:802–809.
148. Zhang H, Ryu D, Wu Y, et al. NAD(+) repletion improves mitochondrial and stem cell function and enhances life span in mice. *Science.* 2016;352:1436–1443.
149. de Guia RM, Agerholm M, Nielsen TS, et al. Aerobic and resistance exercise training reverses age-dependent decline in NAD(+) salvage capacity in human skeletal muscle. *Physiol Rep.* 2019;7:e14139.
150. Fang EF, Kassahun H, Croteau DL, et al. NAD(+) replenishment improves lifespan and healthspan in ataxia telangiectasia models via mitophagy and DNA repair. *Cell Metab.* 2016;24:566–581.
151. Janssens GE, Grevendonk L, Zapata Perez R, et al. Healthy aging and muscle function are positively associated with NAD+ abundance in humans. *Nature Aging.* 2022;2:254–263.
152. Akbari M, Kirkwood TBL, Bohr VA. Mitochondria in the signaling pathways that control longevity and health span. *Ageing Res Rev.* 2019;54:100940.
153. Petersen KF, Befroy D, Dufour S, et al. Mitochondrial dysfunction in the elderly: possible role in insulin resistance. *Science.* 2003;300:1140–1142.
154. Guo C, Sun L, Chen X, Zhang D. Oxidative stress, mitochondrial damage and neurodegenerative diseases. *Neural Regen Res.* 2013;8:2003–2014.
155. Sreedhar A, Aguilera-Aguirre L, Singh KK. Mitochondria in skin health, aging, and disease. *Cell Death Dis.* 2020;11:444.
156. Suski JM, Karkucinska-Wieckowska A, Lebiedzinska M, et al. p66Shc aging protein in control of fibroblasts cell fate. *Int J Mol Sci.* 2011;12:5373–5389.
157. Salminen A, Ojala J, Kaarniranta K, Kauppinen A. Mitochondrial dysfunction and oxidative stress activate inflammasomes: impact on the aging process and age-related diseases. *Cell Mol Life Sci.* 2012;69:2999–3013.
158. Ruetenik A, Barrientos A. Dietary restriction, mitochondrial function and aging: from yeast to humans. *Biochim Biophys Acta.* 2015;1847:1434–1447.
159. Pinton P, Giorgi C, Siviero R, Zecchini E, Rizzuto R. Calcium and apoptosis: ER-mitochondria Ca2+ transfer in the control of apoptosis. *Oncogene.* 2008;27:6407–6418.
160. Sahin E, Colla S, Liesa M, et al. Telomere dysfunction induces metabolic and mito-chondrial compromise. *Nature.* 2011;470:359–365.
161. Wang H, Lautrup S, Caponio D, Zhang J, Fang EF. DNA damage-induced neuro-degeneration in accelerated ageing and Alzheimer's disease. *Int J Mol Sci.* 2021;22.
162. Fang EF, Scheibye-Knudsen M, Chua KF, Mattson MP, Croteau DL, Bohr VA. Nuclear DNA damage signalling to mitochondria in ageing. *Nat Rev Mol Cell Biol.* 2016;17:308–321.
163. Yang S, Lian G. ROS and diseases: role in metabolism and energy supply. *Mol Cell Biochem.* 2020;467:1–12.
164. Turner MD, Nedjai B, Hurst T, Pennington DJ. Cytokines and chemokines: at the crossroads of cell signalling and inflammatory disease. *Biochim Biophys Acta.* 2014;1843:2563–2582.
165. Navarro A, Boveris A. The mitochondrial energy transduction system and the aging process. *Am J Physiol Cell Physiol.* 2007;292:C670–C686.
166. Singh R, Letai A, Sarosiek K. Regulation of apoptosis in health and disease: the balancing act of BCL-2 family proteins. *Nat Rev Mol Cell Biol.* 2019;20:175–193.
167. Criddle DN, Gerasimenko JV, Baumgartner HK, et al. Calcium signalling and pancreatic cell death: apoptosis or necrosis? *Cell Death Differ.* 2007;14:1285–1294.

168. Kania E, Roest G, Vervliet T, Parys JB, Bultynck G. IP3 receptor-mediated calcium signaling and its role in autophagy in cancer. *Front Oncol.* 2017;7:140.
169. Pedriali G, Rimessi A, Sbano L, et al. Regulation of endoplasmic reticulum-mitochondria Ca(2+) transfer and its importance for anti-cancer therapies. *Front Oncol.* 2017;7:180.
170. Reed JC, Jurgensmeier JM, Matsuyama S. Bcl-2 family proteins and mitochondria. *Biochim Biophys Acta.* 1998;1366:127−137.
171. Lin DT, Wu J, Holstein D, et al. Ca2+ signaling, mitochondria and sensitivity to oxidative stress in aging astrocytes. *Neurobiol Aging.* 2007;28:99−111.
172. Mattson MP, Arumugam TV. Hallmarks of brain aging: adaptive and pathological modification by metabolic states. *Cell Metab.* 2018;27:1176−1199.
173. Okin D, Medzhitov R. Evolution of inflammatory diseases. *Curr Biol.* 2012;22:R733−R740.
174. Picca A, Lezza AMS, Leeuwenburgh C, et al. Fueling inflamm-aging through mitochondrial dysfunction: mechanisms and molecular targets. *Int J Mol Sci.* 2017;18.
175. Riley JS, Tait SW. Mitochondrial DNA in inflammation and immunity. *EMBO Rep.* 2020;21:e49799.
176. Shimada K, Crother TR, Karlin J, et al. Oxidized mitochondrial DNA activates the NLRP3 inflammasome during apoptosis. *Immunity.* 2012;36:401−414.
177. Smith JA. STING, the endoplasmic reticulum, and mitochondria: is three a crowd or a conversation? *Front Immunol.* 2020;11:611347.
178. Mittal M, Siddiqui MR, Tran K, Reddy SP, Malik AB. Reactive oxygen species in inflammation and tissue injury. *Antioxid Redox Signal.* 2014;20:1126−1167.

SECTION IV

Senescence

CHAPTER 6

Senescence in aging

Sofie Lautrup[1], Alexander Anisimov[1], Maria Jose Lagartos-Donate[1] and Evandro Fei Fang[1,2]

[1]Department of Clinical Molecular Biology, University of Oslo and Akershus University Hospital, Lørenskog, Norway; [2]The Norwegian Centre on Healthy Ageing (NO-Age) Network, Oslo, Norway

1. What is senescence?

The cellular process of irreversible growth arrest was first described over half a century ago by L. Hayflick in primary eukaryotic cells.[1] Hayflick described a phenomenon of limited cellular proliferation or growth in normal human cells in culture linked to both aging and tumor suppression. This phenomenon is now known to be the process of senescence,[2] a process that can be induced by numerous stressors including telomere shortening, exposure to genotoxic agents, nutrient deprivation, hypoxia, mitochondrial dysfunction, and oncogene activation.[3,4] Although cellular senescence is a hallmark of aging and has been linked to age-related degeneration, it is also essential during normal development and is required for proper tissue homeostasis[5]; in this way, senescence is often viewed to as a double-edged sword.[6] Senescent cells fail to proliferate in response to physiologic signals, also called mitogens, and show selected changes in differentiated functions. Moreover, senescent cells refuse to die via programmed cell death or apoptosis, and they secrete an array of senescence-associated signaling phenotype (SASP) molecules, which can have deleterious effects on the surrounding microenvironment and lead to senescence in neighboring cells. This will be explained in more detail throughout the chapter. These attributes of senescent cells have been demonstrated in mice, in which transplantation of either senescent cells or organs from old mice into young mice led to an increase and spread of the senescent phenotype.[7,8]

1.1 Characteristics of senescent cells

Senescent cells have several features, although no feature is exclusively specific. The various senescence markers depend on the cell type, the stressor inducing senescence, and the microenvironment surrounding the senescent cell (Fig. 6.1). A common feature is irreversible cell cycle arrest,

Molecular, Cellular, and Metabolic Fundamentals of Human Aging
ISBN 978-0-323-91617-2
https://doi.org/10.1016/B978-0-323-91617-2.00010-9

© 2023 Elsevier Inc.
All rights reserved.

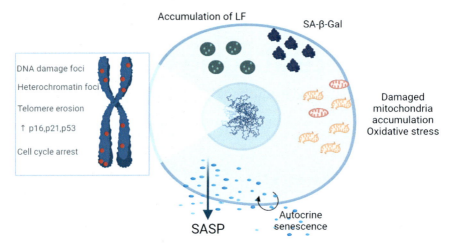

Figure 6.1 *Key features of cellular senescence. Senescent cells display an array of features dependent on senescence inducer, cell type, tissue, and so on. Cytoplasmic features of senescent cells include increased senescence-associated β- galactosidase staining, accumulation of lipofuscin, and increased mitochondrial mass owing to an accumulation of damaged mitochondria, resulting in increased oxidative stress and inflammation. In the nucleus, the senescent cells normally show increased levels of DNA-damaged foci, heterochromatin foci, and telomere erosion. In addition to increased cell cycle inhibitors such as proteins p16 and p21, p53 is often increased, all of which lead to permanent proliferation arrest. In addition, senescent cells often exhibit a senescence-associated secretory phenotype that affects both the cell itself (autocrine) and neighboring cells (paracrine). This figure was created using BioRender.*

although even this is not a sole marker of senescence, occurring as it does during quiescence—a reversible, low-energy state found in inactive cells—or the state of terminal differentiation, in which cells acquire specific cellular functions associated with their state of arrest. The by-products of DNA damage, such as the accumulation of double-strand breaks (DSBs), along with telomere shortening have key roles in senescence-associated cell cycle arrest.[6,9] Genomic instability activates the DNA damage response (DDR), characterized by phosphorylated histone H2AX, 53BP1, MDC1, ataxia–telangiectasia mutated (ATM), and ataxia telangiectasia and Rad3-related protein (ATR), as well as the downstream kinases CHK2 and CHK1.[10] In addition, the expression of essential DNA replication-associated genes is repressed during senescence, contributing to cell cycle arrest (more details are provided subsequently). Second, swelling of both the cell in general and of the nucleus can be a sign of senescence.[1,11] Third, senescent cells exhibit increased activity of the lysosome associated protein β-galactosidase (β-Gal). Senescent cells can therefore be marked by treatment with a nonstained substrate of β-Gal called X-gal. Upon hydrolysis of X-gal by β-Gal, it converts into a blue precipitate, and the senescent cells will turn blue. Fourth, senescent cells with persistent DDR activation secrete an array of growth factors, proteases, cytokines, and other factors, collectively called SASP. The SASP will differ depending on the induction of senescence as well as the cell type, but common to all types is that it will affect the environment around and the cells adjacent to the senescent cell, resulting in a spread of senescence. SASP will be explained in greater detail later in the chapter. Senescent cells can also exhibit nuclear DDR foci on both dysfunctional telomeres and nontelomere regions of the DNA. These foci contain activated DDR proteins (such as phospho-ATM, phospho-ATR, and related substrates) and are known as DNA segments with chromatin alterations reinforcing senescence.[12,13]

1.2 p53 and p16: hallmarks of senescence

Senescence-associated cell cycle arrest largely depends on two major pathways: p53/p21 and p16/pRB.[14] Both p53 and pRB are master transcriptional regulators, and p21 is a downstream effector of p53, whereas p16 is an upstream regulator of pRB (Table 6.1 provides an overview of the genes and proteins described here).[15] Transcription factor p53 regulates several cellular stress responses. Upon activation, it controls the expression of several genes important in cell cycle arrest, DNA repair, apoptosis, and senescence. It is regarded as one of the most powerful tumor suppressors owing to its ability to stall cell proliferation and induce apoptosis. Unsurprisingly, p53 is lost or mutated in most cancers. Therefore considerable research has been done on p53 in anticancer therapies.[16] However, p53 is also involved in development, reproduction, metabolism, and longevity, as shown by studies in mice and more short-lived animals including *Caenorhabditis elegans* and *Drosophila melanogaster*.[16] In mice, the expression of a constitutively active mutant form of p53 as well as p53 overexpression resulted in accelerated aging and reduced life span, compromising healthy aging.[17−19] Combined, findings in mouse models of p53 mutations suggest that physiologic p53 prevents cancer and protects from aging, whereas overexpressed or overactive p53 compromises healthy aging because p53 elicits cell cycle arrest and promotes cellular senescence.[16] Increased activation of p53 correlates with increased senescence, and studies have shown that activation of p53 can induce a senescence-like phenotype in a DDR-dependent or independent way, whereas inactivity or low levels of p53 can

Table 6.1 Gene and protein names of central senescence-associated factors.

	Overview of gene and protein names related to senescence induction			
Gene name (human)	Full gene name	Protein name	Main function(s)	In senescence
CDKN1A	Cyclin-dependent kinase inhibitor 1A	p21	Negative regulator of cell cycle, transcription factor	Increased/activated
CDKN2A	Cyclin-dependent kinase inhibitor 2A	p16	Negative regulator of cell cycle, transcription factor	Increased/activated, prevents Rb phosphorylation
		p14 (alternative reading frame)	Negative regulator of cell cycle	Increased
CDKN2D	Cyclin-dependent kinase inhibitor 2D	p19	Negative regulator of cell cycle	(Increased)
RB1	RB transcriptional corepressor 1	Rb	Negative regulator of cell cycle, tumor suppressor, chromatin stabilization	Hypophosphorylated by p16
TP53	Tumor protein P53	p53	Tumor suppressor, negative regulator of cell cycle, regulates apoptosis, senescence, DNA repair	Increased/activated, phosphorylates p21

delay or even completely halt the development of senescence.[20–22] Telomere damage is recognized as DSBs and activates the DDR.[23] Furthermore, DNA damage, hyperactivation of oncogenes, and inactivation of oncosuppressors owing to replicative arrest activate DDR.[24,25] DDR proteins including ATM and ATR kinases are activated in response to replication stress or DNA damage, and subsequently activate p53 via phosphorylation. p53 then activates its target, p21, an essential component in p53-mediated cell cycle arrest and senescence. Lack of p21 stops senescence within various contexts[26–28]; however, in situations in which p21 is depleted, p53 can still mediate senescence.[29–31] Prolonged culturing of cells resulted in increased activity of p53. In line with this, p53 shows increased phosphorylation, DNA binding, and transcriptional activity in senescent cells. Despite the previous findings showing that increased p53 activity is associated with senescence, it is not feasible simply to inactivate or delete p53. Human cells with deactivated p53 will exhibit an increased replicative life span, but they will end in a state of cellular crisis characterized by massive genomic instability and frequent cell death. This indicates that we cannot solely use p53 as a switch to turn senescence off and on.

When a senescent-like phenotype is triggered in cells that overexpress cell cycle inhibitors such as p16 or p21 but without DNA damage, the cells undergo growth arrest with many characteristics of senescence but not an SASP as seen in p53-induced senescence.[4] p16 is a cyclin/CDK inhibitor encoded by the gene *CDKN2A* in the *INK4a/Alterate reading frame (ARF)* locus. p16 prevents the phosphorylation of Rb by cyclin/CDK complexes. Hypophosphorylated Rb prevents cell cycle progression by binding and thereby inhibiting E2F, resulting in prevention of cell proliferation and DNA replication. The suggested a role of p16-mediated Rb hypophosphorylation in maintaining senescence, which came from observing decreased p53 levels after the induction of senescence coinciding with a high steady-state level of p16. In addition, in cells with a low level of p16, p53 induced replication and growth whereas cells expressing high p16 levels did not.[25,32] The relevance of p16 in age-associated senescence becomes more striking when noting that the expression of p16 correlates with chronologic aging in both humans and mice across tissues.[33,34] Genome-wide association studies showed a strong correlation between the *INK4a/ARF* locus and several age-associated diseases including Alzheimer's disease (AD).[35,36] Combined, these results present the different roles of the p53/p21 axis and the p16/Rb axis in the two waves of senescence: an early reversible phase and a later irreversible phase, respectively.

1.3 SASP and its constituents

The senescent phenotype constitutes more than cell cycle arrest. Because of the continuous metabolic activity and extensive changes in protein expression and secretion, the senescent cells finally develop a phenotype, SASP.[4] SASP is composed of an array of proteins, soluble and insoluble factors secreted by the senescent cell into the extracellular milieu. Several different cell types such as liver stellate cells, endothelial cells, and epithelial cells of the retinal pigment, the mammary glands, the colon, lung, pancreas, and prostate have been shown to develop an SASP, which has been extensively characterized.[4,37] A brief summary of the major constituents of the SASP follows.

SASP may signal cellular damage or dysfunction to the surrounding tissue or cells, including immune cells, and stimulate repair if needed. SASP is at least partially regulated at the transcriptional level, and it has also been suggested that SASP may be regulated at the chromatin organization level rather than by transcription factors.[4] It is known that chromatin alterations occur in senescence.[38,39] Furthermore, it is likely that SASP is irreversible, whereas cell cycle arrest might be reversible.[32,40] Combined, this suggests that SASP as an essential hallmark of senescent cells. Because of our limited knowledge of the SASP, fundamental questions remain. Future studies on how the SASP is activated and maintained and how it affects the microenvironment around the senescent cell in healthy, diseased, and aged conditions are needed.

The constituents of SASP vary among cell types and treatments. In general, the components of SASP differ from the cause of senescence and in cell types, tissues, and species; however, some factors in SASP are more general. Table 6.2 gives a general overview of SASP and its relation to p53 and oncogene activation, and to mitochondrial dysfunction–induced senescence (MiDAS). Overall, SASP is composed of four classes of factors that are changed during senescence: (1) soluble signaling factors, (2) extracellular proteases, (3) extracellular insoluble factors, and (4) nonprotein secretions.

1.3.1 Proinflammatory interleukins and cytokines

Senescent cells secrete soluble signaling factors often related to the response of immune cells to an infection. However, during senescence, nonimmune

Senescence in aging   155

Table 6.2 Senescence-associated signaling phenotype (SASP) parameters under different conditions.

	Senescence inducers			
SASP factors	General secretory profile	DNA damage	Loss of p53 and/or Oncogenic Ras	Mitochondrial dysfunction
Interleukins (ILs)				
IL–6	↑	↑	↑	↑
IL–7	↑		↑	
IL–1α, –1β	↑	↑	↑	↑
IL–13	↑		↑	
IL–15	↑		↑	
Chemokines				
IL–8	↑	↑	↑	↑
GROα, β	↑	↑	↑	↑
MCP-2	↑		↑	
MCP-4	↑		X	
MIP-1α	↑	↑	↑	
MIP-3α	↑		↑ or x	
HCC-4	↑		X	
Ecotaxin-3	↑		↑	
TECK	X		↑	
ENA-78	X		↑	
I-309	X		↑	
I-TAC	X			
Other inflammatory factors				
Granulocyte macrophage-colony stimulating factor	↑		↑	
Granulocyte-colony stimulating factor	↑ or x		↑	
Interferon-gamma	X		↑	
BLC	X		↑	
MIF	↑		↓	
GCP-2			↑ (x in fibroblasts)	
ICAM-1		↑		

Continued

156 Molecular, Cellular, and Metabolic Fundamentals of Human Aging

Table 6.2 Senescence-associated signaling phenotype (SASP) parameters under different conditions.—cont'd

Senescence inducers				
Lamin B1	↓			↓
P53	↓			↑
AMPK	↓			↑
Nuclear factor-κB	↑		↑	↓
NAD⁺/NADH	↑			↓
Growth factors and regulators				
Amphiregulin	↑		X	
Epiregulin	↑		X	
Heregulin	↑		X	
Epidermal growth factor	↑ or x		X	
Insulin growth factor (IGF) receptor network (fibroblasts, endothelial, and epithelial)				
IGFBP-2	↑	↑		
IGFBP-3	↑	↑		
IGFBP-4	↑			
IGFBP-5	↑			
IGFBP-6	↑			
IGFBP-7			↑	

related cells additionally secrete factors such as interleukins, inflammatory cytokines, and growth factors, which can affect the extracellular environment and neighboring cells.[4]

Inflammatory interleukins (ILs), such as IL-6, IL-1α, and IL-1β, form most proinflammatory secretions, along with some colony stimulating factors (CSFs) (such as granulocyte macrophage [GM]-CSF and granulocyte [G]-CSF), among others. IL-6 is the most prominent cytokine of SASP. It is proinflammatory and associated with DNA damage and oncogenic stress-induced senescence in both mouse and human keratinocytes, melanocytes, monocytes, fibroblasts, and epithelial cells.[14,41–43].

(i) IL-6 have been associated with oncogene stress-induced senescence, as well as persistent DNA damage. IL-6 secretion has also been linked to DNA damage signaling via ATM and CHK2, but is independent of the p53 pathway.[9,37,41,42] In addition, senescent cells secreting IL-6 can directly affect neighboring cells expressing IL-6 receptors (gp80 and gp130).

(ii) IL-1α is central in IL signaling and is part of the early response to senescence-inducing stimuli. This plasma-bound IL binds to its receptor (IL-1R), which then activates a signaling cascade converging in nuclear factor-κB (NF-κB) activation. The transcription factor NF-κB again stimulates the expression of an array of inflammatory mediators such as IL-6.[44,45] For instance, senescent fibroblasts overexpress and secrete both IL-1α and IL-1β,[46] and both interleukins can affect neighboring cells via cell surface receptors including IL-1R, which then triggers NF-κB and activator protein 1 pathways.[47]

(iii) Other cytokines including CSFs (e.g., GM-CSF and G-CSF) and osteoprotegerin are also increased during senescence, especially in fibroblasts.[9,14,42,48] Moreover, specific secretory molecules including IL-2, IL-4, IL-10, IL-11, and IL-12 are not changed during senescence in mouse fibroblasts under physiologic 3% oxygen,[4,14] which exemplifies some of the differences between the SASP and the secretory profile during an immune response.

1.3.2 Chemokines

Most senescent cells overexpress CXCL-8 (IL-8), CXCL-1, and CXCL-2 (GROα and GROβ, respectively). In the C−C motif ligand (CCL) family, MCP-2, -4, and -1, HCC-4, eotaxin-3, and MIP-3α and -1α are among often upregulated genes.[4]

1.3.3 Insulin-like growth factor−binding proteins

The insulin-like growth factor (IGF) pathway also affects the microenvironment surrounding the senescent cells. Senescent endothelial, epithelial, and fibroblast cells express high levels of multiple IGF-binding proteins (IGFBP-2 to -6).[14,49−51] It has been shown that secretion of IGFBP-7 by senescent primary fibroblast, for example, can lead to senescent and apoptosis in neighboring cells.[51]

1.3.4 Extracellular proteases

SASP proteases can have three major effects: (1) shedding of membrane-associated proteins, resulting in soluble versions of membrane-bound receptors; (2) cleavage or degradation of signaling molecules; and/or (3) degradation or processing of the extracellular matrix (ECM). These activities provide potent mechanisms by which senescent cells can modify the tissue microenvironment.[4]

(i) Matrix metalloproteases

Matrix metalloprotease (MMP)-1, MMP-3, and MMP-10 are steadily increased in human and mouse fibroblasts undergoing replicative or stress-induced senescence. In some instances, these MMPs can regulate some of the soluble SASP factors. For instance, MMP-1 and MMP-3 produced by senescent cells can cleave MCP-1, -2, and -4, as well as IL-8, participating in regulating SASP.[52,53] In addition, inactive forms of CXCL- and CCL-family chemokines released from neighboring cells such as tumor cells or leukocytes can be processed by MMP-2, 7, and -9.[54]

(ii) Serine proteases and regulators of the plasminogen activation pathway

Members of the plasminogen activation (PA) pathway, including urokinase-type PA (uPA) or tissue-type PA (tPA), the uPA receptor, and PA inhibitors (PAI-1 and -2) have been demonstrated as part of SASP in endothelial, lung, and skin fibroblasts.[55] Plasminogen activity has been reporter to increase more than 50-fold in senescent fibroblasts from lung and skin, and in senescent endothelial cells. In addition, PAI-1 is increased in senescent fibroblasts and endothelial cells from aged human donors, and PAI-1 seems to reinforce senescence-induced growth arrest.[56]

1.3.5 Fibronectin

The ECM has complex functions in mechanical support to both the cells and tissues, but it has also been shown to exhibit more physiologic roles. For example, the extracellular insoluble glycoprotein fibronectin works as storage for various growth factors, and thus is involved in regulating cell growth, morphogenesis, and inflammatory responses. It has been shown that the composition of the ECM, including the levels and stretching of fibronectin, changes during aging.[57] Studies revealed that fibroblasts from patients with accelerated aging diseases such as Werner syndrome, as well as senescent cells both in vitro and in vivo have increased levels of fibronectin.[58,59]

1.3.6 Nonprotein secretion

Senescent cells exhibit altered cellular metabolism leading to the secretion of nonprotein signaling molecules such as reactive oxygen species (ROS), nitric oxide, and various transported ions. These secreted molecules alter the surrounding microenvironment, affecting both the secreting cell and its neighbors.[60,61] Reactive molecules such as ROS are known to affect the ability and pattern of differentiation of monocytes, for example. They can also enhance cancer aggressiveness and are believed to promote aging and age-related diseases, as described in the free radical theory of aging.[62]

2. Current concepts of senescence in aging

Normal cells do not proliferate and divide indefinitely because the process of senescence limits their replicative life span or their capacity for cell division. Senescence has been presented as both a pathway that suppresses tumorigenesis and one that contributes to aging, at least in mitotic tissues, based on the finding that the number of senescent cells increases with age.

Fibroblasts from older individuals have been shown to senesce after fewer generations or cell divisions compared with fibroblasts from younger individuals, which suggests that mitotic cells lose replicative potential with age.[13,63,64] Moreover, primary fibroblasts from patients with the progeroid disease Werner syndrome go into senescence much earlier than do cells from normal age- and sex-matched individuals.[65,66] In contrast to cells from many rodent species, human cells rarely spontaneously immortalize.[67,68] This means that senescence is stringent, and only a few cell types might not experience senescence, likely owing to an indefinite replicative life span. For example, certain stem cells, germ line cells, and tumor cells exhibit unlimited replication and therefore will not senesce normally. It remains uncertain whether specific types of stem cells may divide indefinitely.

During aging, the number of senescent cells increases. This suggests that senescence might be a beneficial compensatory response that prevents tissues from damage. However, cells further require efficient mechanisms to replace and clear the senescent cells. In aged organisms, the turnover system may become inefficient or may exhaust the regenerative capacity of progenitor cells. The loss of these systems may eventually result in accumulation of senescent cells, which might exacerbate the damage and contribute to the aging process.[5]

2.1 Mouse models revealing the impact of senescence on aging

Obviously, humans are not the only organism for whom senescence has been studied. Studies on mouse models have had an important role in understanding the role of senescence in aging.[69] Clearance of p16-positive senescent cells alleviates the onset of several age-related pathologies in progeroid and wild-type mice.[70,71] The involvement of aging in multiple diseases has been studied using mouse models, and there is evidence that senescence causes multiple chronic conditions such as neurodegeneration, myocardial infarction, and diabetes, among others. In addition, cultured senescent mouse fibroblasts under physiologic 3% oxygen exhibit an SASP comparable to human cells.[14] Mice exhibit extremely long telomeres, which might explain some of the differences seen in senescence studies of mice versus human models. Mouse studies in which old mice receive a younger heart transplant, or young mice receive an older heart transplant, also support the age-related increase in senescence. It was shown that young mice with an old heart had an increased amount of senescent cells.[8] In addition, mice that were transplanted with senescent cells showed increased physical dysfunction and immune system dysfunction, and increased senescence.[7]

Different mouse models have been developed to show that the accumulation of senescent cells leads to age-related diseases and aging (Table 6.3 provides an overview of the mouse models described subsequently).[71,75,76,77] In a progeroid mouse model expressing 10% of the kinase BubR1 ($BubR1^{H/H}$), involved in cell cycle regulation, the mice developed age-related senescence in several tissues including adipose, skeletal muscle, and eye.[76,77] The transgene INK-ATTAC consists of a fragment of the p16 promoter active only in senescent cells, followed by green fluorescent protein (GFP) to allow for visualization of the senescent cells, and a drug-sensitive protein. This enables the selective killing of senescent p16-expressing cells by drug treatment. Expression of the INK-ATTAC transgene in the $BubR1^{H/H}$ mouse model was used to demonstrate that removing p16-expressing senescent cells from weaning age onward attenuated the accumulation of senescent cells, preventing premature aging.[71] A second study using the INK-ATTAC model in naturally aged mice corroborated these findings,[70] and the mice had an increased median life span and increased health span.[3,70]

The importance of p16 in senescence was further demonstrated in a senescence reporter mouse model. The p16-3MR mouse model expresses a trimodal reporter fusion protein under the control of an artificial promoter for p16. This leads to p16-expressing cells becoming sensitive to elimination by the drug ganciclovir (a nucleoside analogue that is converted into a toxic DNA chain terminator by herpes simplex virus thymidine kinase, causing cell death).[78] However, it remains unclear whether the elimination of the senescent cells themselves or elimination of the SASP is the key element underlying these effects.[3]

Table 6.3 Overview of mouse models described in this chapter.

Mouse models found in this chapter

Model	Transgene	Molecular consequence	Phenotypes related senescence and aging	References
Alzheimer's disease mouse models				
rTg(tauP301L) 4510	Overexpression of human tau protein with P301L mutation	Mice exhibit well-characterized, aggressive tau pathology in forebrain regions concomitant with neurodegeneration and cognitive deficits.	Increased senescence: Increased expression of _Cdkn1a_ and _Cdkn2a_	72,73
MAPT P301S PS19	Overexpression of mutant human tau specifically in neurons	Mice exhibit gliosis, neurofibrillary tangle deposition, neurodegeneration, and decreased cognitive function.	Increased senescence: Accumulation of p16-positive senescent astrocytes and microglia	74
Aging				
BubR1^{H/H}	Hypomorphic allele of BubR1	90% reduction in BubR1 expression compared with wild-type mice	Progeroid mice: Decreased life span Premature age-related phenotypes. Accumulation of p16-positive cells in certain tissues (adipose tissue, skeletal muscle, and eye)	71,75,76 68,69,71

(Continued)

BubR1$^{H/H}$ INK-ATTAC	Expression of transgene INK-ATTAC in *BubR1$^{H/H}$* background	Expression of INK-ATTAC allows for inducible elimination of p16-positive senescent cells upon administration of drug	Removal of p16-expressing cells prolonged life span of and delayed tissue degeneration in progeroid BubR1 mice
DNA PolG	Mouse model with DNA polymerase gamma mutation (POLGD257A)	Mutation in proofreading domain of mitochondrial DNA polymerase (DNA PolG)	Accelerated aging, rapid accumulation of mitochondrial DNA mutations.

Senescence reporter

P16-3 MR	Expresses trimodal reporter fusion protein under control of artificial promoter for p16	p16-expressing cells become sensitive to elimination by ganciclovir	Senescent cells expressing p16 can be removed. [77]

Although senescence accelerates aging, it might also protect health. For example, senescence inhibits cancer/tumor formation in addition to reducing fibrotic scars.[79] Abrogation of senescence might be fundamental to tumor formation, but this protective effect of senescence might fail in the later stages of life, resulting in the accumulation of senescent cells.[80]

2.1.1 Senescence and its links to age-related diseases

Senescence has been linked not only to the process of aging but also to a variety of age-related diseases including atherosclerosis,[81,82] pulmonary fibrosis,[83] hepatic steatosis,[84] and neurodegenerative diseases. A more detailed overview of the findings of senescent cells in dementia will be presented subsequently.

2.2 Senescence and neurodegeneration

Multiple studies have demonstrated senescence in mitotic cells in the central nervous system, such as microglia, astrocytes, and progenitor cells, but neurons might also enter senescence or senescence-like states.[85]

Microglia and astrocytes are the main glial cells of the brain and serve as the primary immune response of the brain. They become activated as a response to pathogenic invasion or injury, initiating a proinflammatory response. Astrocytes are also essential to the blood—brain barrier and have a central role in providing energy to neurons via the astrocyte—neuron lactate shuttle.[86] Various markers of senescence have been observed in patients with neurodegenerative diseases, but the role of senescence in the etiology of these diseases is still largely unexplored. AD is the most common type of dementia. The 2021 annual report from the World Health Organization estimated that around 55 million people worldwide were affected.[87] The traditional hallmarks of AD include aggregated β-amyloid (Aβ) plaques and neurofibrillary tangles (NFTs) consisting of hyperphosphorylated Tau protein, as well as severe neuroinflammation, neuronal degeneration, and brain atrophy. Studies have demonstrated new causes and risk factors, such as impaired cellular metabolism, accumulated damaged mitochondria, and compromised mitophagy.[88,89] Finally, AD patients exhibit cognitive and memory impairments associated with neuronal loss and dysfunction.

A central role of senescence in AD is supported by multiple studies in both AD mouse models, human postmortem brain samples, and in vitro cell cultures. However, the link between senescence and Tau and/or Aβ pathology remains unclear.

In mice, Tau-based and Aβ-based pathology have been linked to senescence in different models. Tau NFTs have been associated with dysfunctional mitochondria and increased expression of *Cdkn1a* and *Cdkn2a* as a measure of senescence in the rTg(tauP301L)4510 transgenic mouse model of AD,[72] and accumulation of p16-positive senescent astrocytes and microglia was found in the microtubule-associated protein tau (MAPT) P301S PS19 model.[74] Clearance of p16-positive cells prevented gliosis, the response of reactive astrocytes to brain injury or insult, hyperphosphorylation of both soluble and insoluble tau—decreasing NFT deposition—and degeneration of cortical and hippocampal neurons, leading to preserved cognitive function.[72,90] The link between AD and senescence has been further validated by transcriptomic analysis of human brain regions with severe Tau pathology. These analyses indicated that cortical neurons with NFTs showed increased senescence compared with control (non–NFT expressing) cortical neurons. This was also shown in another tauopathy, progressive supranuclear palsy, in which brain samples showed increased expression of *CDKN2A* compared with controls.[72]

Aβ pathology has also been linked to senescence in the APP/PS1 AD mouse model, which showed significant increased p16 mRNA puncta and senescence-associated β-Gal (SA-β-Gal)-positive cells associated with Aβ plaques in the entorhinal cortex and hippocampus, but not in the triple transgenic AD mouse model.[72,90] In vitro Aβ treatment of human astrocytes in culture triggered senescence, as measured by increased p16, SA-β-Gal, and IL-6.[91] Moreover, increased levels of p16 and certain metalloproteases (MMP-1 and MMP-3) have been observed with increased age in human postmortem brain samples.[91] Oligodendrocyte progenitor cells (OPC) from control, mildly cognitively impaired, and AD patients showed differences with regard to senescence markers *CDKN1A* and *CDKN2A*; almost all OPCs were positive for Aβ and the two senescence markers in the AD brain tissues.[74,90] Finally, microglia from AD patients have shorter telomeres,[92] and astrocytes show several markers of SASP (including p16, MMP-1, and IL-6)[91] and increased DNA damage.[93]

The underlying mechanisms and role(s) of senescence in AD and other neurodegenerative diseases remain elusive. Further research is needed to understand the extent of senescence, aging, and neurodegeneration in the brain. The future will tell whether targeting senescent cells might form a new interventional strategy against AD.

3. Factors driving and/or triggering senescence in aging

Aging has been linked to an increase in senescence, likely owing to an overall increased level of various stresses related to senescence. These stresses include genomic instability caused by extensive DNA damage and chromatin disruption, telomere dysfunction as a result of repeated cell division (known as replicative senescence), mitochondrial damage and dysfunction, and expression of certain oncogenes.[4]

As opposed to age-related or chronic senescence, disposal of senescent cells during wound healing, tissue repair, and embryogenesis is necessary for the function of the organism. It is characterized by efficient and strict temporal control,[2,94−96] called acute senescence, and it allows for a targeted effect on a specific population of cells.[97] The mechanism of acute senescence involves a rapid response to and repair of genomic damage.[98] However, a crucial role in establishing an aging-related senescent phenotype is continuous macromolecular damage, resulting in persistent cell cycle arrest and the development of the SASP.[97,99] In the next section, we will focus on some of the main pathways that can induce senescence: genomic instability, telomere attrition, mitochondrial dysfunction, and oncogene-induction (Fig. 6.2). In addition, we will provide a brief overview of the links between senescence and age-related stem cell exhaustion.

3.1 Genomic instability as a trigger and sign of cellular senescence

Genomic instability increases during aging owing to both increased DNA damage and inefficient repair mechanisms. Genomic instability can induce DDR, and senescence can be triggered by severely damaged DNA and prolonged DDR activation. Persistent DDR activation is generally characterized by the continuous presence of DNA damage foci, which contain an array of DDR-related proteins, including phosphorylated and thus activated p53.[15] DDR uses specific checkpoint proteins to block progression of the cell cycle and prevent the propagation of mutated genetic information to daughter cells.[3] The DDR machinery, including the kinase ATM and ATR, is activated during the early stages of senescence. Activates ATM and ATR can phosphorylate and then activate p53. Phosphorylated p53 can then induce the expression of p21, which can further regulate and block cell cycle. In the following state of senescence, the expression of p16 is enhanced, likely to maintain the senescent phenotype including

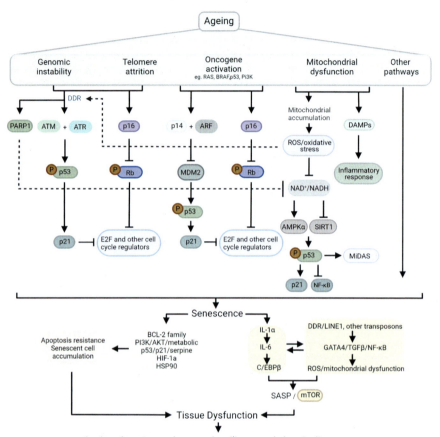

Figure 6.2 *Mechanisms of age-dependent senescence. The four main inducers of age-related senescence are genomic instability, telomere attrition, oncogene activation, and mitochondrial dysfunction. Both genomic instability and telomere shortening or attrition can activate the DNA damage response (DDR), which induces cell cycle arrest via ataxia-telangiectasia mutated (ATM)-ataxia telangiectasia and Rad3-related (ATR)-p53-p21 or p16-Rb-E2F, respectively. Oncogene-induced senescence (OIS), involving Rat sarcoma (RAS), Rapidly accelerated fibrosarcoma homolog B (BRAF), and p53 among others, is activated owing to the activation of p14^{ARF} inhibiting MDM2. This leads to the phosphorylation and activation of p53 and then p21. OIS is also occurring via increased p16 expression and subsequent Rb phosphorylation (inhibition), inhibiting E2F and other cell cycle regulators, as seen in DDR and telomere attrition-induced senescence. Finally, mitochondrial dysfunction and associated mitochondrial accumulation increase the level of reactive oxygen species (ROS) and thus oxidative stress. In turn, NAD$^+$ is depleted and the energy sensor AMPKα activated. AMPKα activates p53, which activates p21, leading to cell cycle arrest. Dysfunctional mitochondria also produce damage-associated molecular patterns (DAMPs) such as ROS and cardiolipin. All of these pathways can lead to senescence. Senescence can then activate an*

inflammatory response, which, in addition to DDR and mitochondrial dysfunction, can lead to the generation of a senescence-associated signaling phenotype (SASP) and/or increased mTOR. The SASP can act in an autocrine way to inhibit cell cycle, among others, but also in a paracrine way, in which the secreted molecules affect the surrounding environment. Senescence also affects apoptosis-related proteins such as the BCL-2 family, in addition to phosphatidylinositol 3-kinase (PI3K)/AKT and p53/p21/ Serpine[100] pathways, HIF-1α and HSP90, which leads to resistance to apoptosis, and then senescent cells accumulate. In total, this will lead to tissue dysfunctions, again resulting in aging phenotypes, decreased resilience, and an array of diseases. Other pathways, both known and likely also unknown, will be able to induce senescence and the senescent phenotype. The figure is based on the literature and the ideas of J. Kirkland, Mayo Clinic, United States. HSP90, heat shock protein 90; MiDAS, mitochondrial dysfunction-induced senescence; NF-κB, nuclear factor-κB; SIRT1, sirtuin 1.

inhibition of cell proliferation.[101] Although p16 is a central marker of senescence, p16 induction alone is not sufficient to induce a complete SASP.[101]

Activation of DDR in senescent cells can result in SASP, although not in all cases. Normally, DDR is activated by DNA damage. For DNA DSBs, repair pathways called homolog recombination or nonhomologous end joining are responsible for repairing damage using the kinase ATM and the MRN complex (consisting of Mre11, Rad50 and NBS1 [Nijmegen breakage syndrome 1]), among others. However, for DDR activation to induce an SASP, persistent activation caused by massive genomic damage, and an accumulation of ATM and NBS1 (part of the MRN complex) on the DNA are required. Furthermore, the expression of CHK2 (checkpoint kinase 2), a cell cycle checkpoint protein, is required.[9] It has been shown that activated ATM, CHK2, and NBS1, but not the expression of p53, are essential for establishing and maintaining an SASP, exemplified by increased secretion of IL-6 and IL-8.[9] This suggests that DDR-mediated regulation of SASP may be one of the routes by which the DDR drives age-related inflammation.[3]

3.2 Telomere attrition

One of the first and best-characterized mechanisms of cellular senescence induction is telomere shortening. Telomeres are repetitive DNA structures found in the terminal loops at the ends of chromosomes and stabilized by the shelterin protein complex. Because the standard DNA replication apparatus does not duplicate the terminal portion of the polynucleotide chains, in the absence of telomere maintenance mechanisms such as the expression of telomerase or recombination among telomeres, telomeres shorten with each round of DNA replication. Below a certain length, the loss of telomere-capping factors or protective structures makes critically short telomeres resemble DSBs and thus triggers a DDR that is similar to the one triggered by DNA DSBs.[102,103] One or a few DDR signaling telomeres are sufficient to trigger replicative cellular senescence.[104] However, forced expression of telomerase can prevent this type of cellular senescence and promote unlimited cell proliferation.[105] This indicates that telomere shortening is most likely involved in organismal aging, which is further reinforced by findings showing that defects in telomere maintenance accelerate aging in mice and humans.[5]

Telomere shortening during cell proliferation culminates in destabilized telomere DNA loops, leading to telomere unpacking, generating telomere dysfunction-induced foci that activate DDR, ultimately causing cell cycle arrest.[106] Another form of DNA damage, called telomere-associated foci, may exist on telomeres owing to oxidative DNA damage in telomere G—rich regions, regardless of telomere length or loss of shelterin[107]. Persistent DDR activation can also occur at telomeres that are not critically shortened, such as in nondividing cells exposed to genotoxic treatments and in older nondividing cells with inefficient DNA repair.[108–110] As telomeric DSBs persist in nonproliferating cells, continued DDR activation at the telomeres triggers and maintains cellular senescence, independent of telomere length.[3,111]

3.3 Compromised mitophagy and impaired mitochondrial function

Mitochondria are the powerhouses of the cell. Therefore, the state of the mitochondria is central to both cellular and organismal health. During aging and in age-related diseases, mitochondrial function decreases and dysfunctional mitochondria accumulate. This process has been highly documented in tissues composed largely of postmitotic cells (e.g., muscle and brain).[112–114] Various processes can cause mitochondrial dysfunction, including extensive formation and decreased removal of ROS, disruption of the antioxidant defense system, and violation of mitochondrial quality control, including the degradation and recycling system, termed mitophagy. A consequence of mitochondrial dysfunction is cellular senescence.[115–117] Senescent cells have increased mitochondrial mass and abundant tricarboxylic acid cycle metabolites.[118,119]

In addition to oxidative stress and ROS, mitochondrial damage-associated molecular patterns (DAMPs) might act as potential stimulators of the SASP. An emerging concept is that inflammation during aging and age-related diseases may appear as a consequence of the innate immune system identifying damaged components of mitochondria acting as DAMPs. Cardiolipin, a glycerophospholipid that resides in the inner mitochondrial membrane to maintain the mitochondrial structure, has been identified as a mitochondrial DAMP and a contributor to mitophagy.[120] The release of oxidized cardiolipin from damaged mitochondria in immune-related cell types such as monocytes/macrophages and neutrophils, led to increased intracellular Ca^{2+} levels, which in turn induced the expression of the endothelial activator leukotriene B4.[121] This exemplifies the interconnectedness of processes controlling how cells react to various stressors, both during normal functioning and in aging.

3.3.1 Mitochondrial dysfunction-induced senescence phenotype

MiDAS exhibits a distinct secretory phenotype that can influence the differentiation of certain cell types.[122] In vitro studies in primary human fibroblasts showed that knocking down mitochondrial deacetylase sirtuin (SIRT)3, and to a lesser extent SIRT5, but not other SIRTS, induced senescence, as confirmed by increased SA-β-Gal staining.[122] Further research showed that reduced lamin B1[123] and decreased nuclear localization of high mobility group protein B1 were also indicators of senescence.[124] In addition, when depleting cells of mitochondrial DNA, inhibiting the electron transport chain, or depleting cellular heat shock 70-kDa protein 9, senescence was induced, again with a secretory phenotype, but different from that of the SASP induced by genotoxic stress described earlier, and independent of ROS and DNA damage. The $NAD^+/NADH$ ratio mediates the MiDAS phenotype, which links it directly back to the importance of mitochondrial function in senescence induction. Because mitochondria oxidize NADH to NAD^+, mitochondrial dysfunction leads to a reduced $NAD^+/NADH$ ratio. Whereas mitochondria oxidize NADH generated by the tricarboxylic acid cycle or fatty acid oxidation, they also oxidize the cytosolic $NAD^+/NADH$ pool through α-glycerophosphate and malate-aspartate shuttles.[125] Inhibition of the latter by depletion of the malate dehydrogenase, which is located in the inner mitochondrial membrane, where it reduces oxaloacetate to malate by oxidizing NADH to NAD^+, lowers the $NAD^+/NADH$ ratio and induces senescence.[126] This suggests that elevated cytoplasmic NADH can drive cells into senescence.

Notably, NAD^+ declines with age in several tissues,[127–130] linking NAD^+ to both senescence and aging. In addition to changes in $NAD^+/NADH$, the MiDAS phenotype involves increased activation of the energy sensor AMPKα (phosphorylation at Thr-172) and increased Ser-15 phosphorylation of p53, mediated by AMPKα. In addition, the MiDAS phenotype displays similarities to the genotoxin-induced SASP, including increased IL-10, tumor necrosis factor-α, CCL27, p21, and p16, but it lacks the IL-1 signaling arm including IL-1, IL-6, and NF-κB (Table 6.1). This was shown by in vivo studies of the DNA polymerase gamma mouse model, which has a mutation in the proofreading domain of the DNA polymerase, resulting in rapid accumulation of mutations in the mitochondrial DNA.[122] Secretion from MiDAS cells has also been shown to inhibit adipogenesis and promote keratinocyte differentiation, which suggests that MiDAS also has paracrine effects.[122]

3.3.2 Macroautophagy and senescence

Dysfunctional cellular organelles are usually degraded through the activation of an intracellular degradation system called autophagy.[131] It is generally believed that autophagy inhibits senescence. Autophagy protected against oxidative stress–induced senescence; enhancing autophagic activity under excessive oxidative stress by mTOR inhibition delayed cellular senescence and functionally restored both mitochondrial and lysosomal functions.[132] During senescence, inhibition of GATA4, a transcription factor for autophagic degradation, was shown to contribute to senescence and SASP.[133] Knockdown of central autophagy proteins ATG7 and ATG12 (autophagic proteins autophagy related) and lysosomal-associated membrane protein 2 also caused a senescence-like state with increased ROS.[134] Treatment of senescent cells with rapamycin, a known autophagy activator, reduced mitochondrial mass and ROS in addition to other markers of cellular senescence both in vitro and in vivo,[135,136] corroborating the role of autophagy in senescence. Accordingly, ROS contribute to the persistence of the DDR in senescent cells, adding to senescence-associated cell cycle arrest.[117]

3.3.3 Mitophagy and senescence

Mitophagy is a specialized form of autophagy in which damaged and dysfunctional mitochondria can be selectively targeted for degradation and recycling. Detailed mechanisms of mitophagy have been summarized elsewhere.[137,138] Because accumulation of damaged mitochondria is a source of MiDAS (as explained earlier), induction of mitophagy, and thus cellular clearance of damaged mitochondria, might inhibit senescence. Experimental evidence is still needed, but some findings support the involvement of mitophagy. The protein parkin is involved in the coordination of mitochondrial degradation. Interestingly, the senescence regulator p53 can suppress mitophagy by interacting with parkin and blocking its translocation to dysfunctional mitochondria, inhibiting mitophagy-mediated degradation.[139] In addition, the prevalence and induction of mitophagy are impaired during senescence in both in vitro and in vivo studies.[140] Moreover, p53-mediated inhibition of mitophagy promotes cardiac dysfunction during aging in vivo, which was counteracted by either overexpression of parkin or deletion of p53.[140] Finally, it has also been suggested that ROS produced in senescent cells can result in the accumulation of p53 in the cytosol where it inhibits parkin, resulting in mitochondrial dysfunction.[141] Further studies on the effects of pharmacologic and genetic upregulation of mitophagy on senescence in different cells, tissues, and species should be prioritized.

3.4 Oncogene-induced senescence and activation of tumor suppressors

Oncogenic activation, which can stimulate cell proliferation, is an underlying mechanism of cancer development and in tumorigenesis in many cancer types. However, it may also lead to oncogenic stress and/or excessive mitogenic signaling, which can stimulate senescence, known as oncogene-induced senescence (OIS); OIS leads to cell cycle arrest and SASP development (Table 6.2 shows the OIS-SASP profile). The permanence of the senescence associated growth arrest enforces the idea that senescence developed, at least in part, as an attempt to suppress the development of cancer and continued tumorigenesis.[11,68,142] Despite the initial belief that OIS protects the organism from cancer and malignancies, subsequent findings suggest that senescence might be deleterious and compromise the effect of chemotherapy.[143] These opposing effects of OIS are likely a consequence of the interactions among aging cells and the microenvironment, inflammation associated with aging, and the SASP formation.[144]

As described in the beginning of this chapter, two of the most important intrinsic cellular regulators of senescence, and therefore also of OIS, are tumor suppressors p53 and p16.[144,145] Oncogenic activation results in excessive mitogenic stimulation, which causes DNA replication stress and DNA damage, including DSBs.[146,147] This, in turn, leads to a persistent DDR, which again converges upon activation of the p53/p21 pathway-mediated cell cycle arrest.[116,147,148]

Several stimuli can induce OIS, such as various forms of mitogenic stimulation, chronic cytokine stimulation, telomere shortening owing to hyperactive cell proliferation, and overexpression of certain growth factor receptors.[119,149–151] Here, one stimulus of OIS and one of tumor suppressor-induced senescence will be presented. The first report of OIS showed that an oncogenic form of H-RAS (H-RAS[V12]), which chronically stimulates mitogen-activated protein kinase (MAPK) signaling pathways, provoked senescence in otherwise normal cells.[142] In line with this finding, several other components from the MAPK pathway can induce senescence when either overexpressed or present in oncogenic forms.[142,152]

Loss or inactivation of the tumor suppressor PTEN has been shown to result in growth factor truncation and induction of p53-dependent senescence. PTEN catalyzes the conversion of the membrane lipid second messenger PIP3 to PIP2 and therefore is a key mediator of the phosphatidylinositol 3-kinase (PI3K) (generates PIP3)-protein kinase B (AKT, PKB) pathway, which has pivotal roles in fundamental cellular functions such as cell proliferation and survival.[153] Furthermore, alteration of the PI3K/AKT pathway is fundamental in many cancers.[154] Up to 70% of primary prostate tumors lose one allele of *PTEN*, so the level of PTEN is critical for the development of prostate cancer. When *PTEN* heterozygosity results in tumor initiation and proliferation, complete loss of *PTEN* inhibits prostate tumorigenesis by inducing p53-dependent senescence.[155] Loss of *PTEN* results in increased activation of p53, which again results in p21-mediated cell cycle arrest and the development of senescence both in vitro and in vivo in mouse models.[155] Therefore, PTEN and its related pathways (e.g., PI3K/AKT) have been studied as a target in anticancer therapy. In preclinical trials, it was demonstrated that treatment with the mTOR kinase 1 complex inhibitor rapamycin had an effect on cancer cells with activated PI3K/AKT signaling.[156,157] The effect of rapamycin is at least partially due to acute PTEN inactivation, which activates p53-dependent senescence in the cancer cells.[151] As opposed to oncogenic H-RAS[V12]-induced senescence, PI3K/AKT-PTEN—associated senescence is independent of DNA damage or persistent DDR as well as hyperproliferation, but dependent on p53.[151] Altogether, these data argue for the heterogeneity of senescence and demonstrate the need for further studies into the causes, triggers, and underlying mechanisms of senescence in disease such as cancer as well as aging.

4. Stem cell exhaustion in aging

A nondividing cell with the ability to proliferate is in a quiescent state. The state of quiescence in stem cells is critical for regulating tissue regeneration, because it helps to preserve key functional characteristics.[158] Undifferentiated stem cells or quiescent cells form a pool for the production, maintenance, and repair of various tissues in the body.[5] Quiescent stem cells have several defense mechanisms that allow them to maintain properties of self-renewal and generation of differentiated cells, which is described as stemness.[159] Stem and progenitor cells are important to the maintenance of tissue homeostasis and organization during physiologic turnover and after tissue and organ injury. Stem cells are extremely vulnerable, so tissue homeostasis can be disrupted by various stressors, including changes in the DNA structure, nonoptimal epigenetic configuration, ROS, telomere attrition, inflammation, mechanical stress, and radiation.[160]

The probability of cellular damage, including DNA damage, and the activation of DDR proteins in stem cells increase as the organism lives, with stem cells becoming more vulnerable with age. Disruption or dysregulation of normal cellular processes leads to decreased regenerative potential in various tissues, which is a of the main characteristic of aging.[161] The functional capacity of stem cells declines with age, which affects the regenerative capacity of tissues during aging.[162] This is caused by a decrease in stem cell activity and terminal cell differentiation as well as by processes that expose stem cells to apoptosis or senescence.[161] Unlike apoptotic cells, senescent stem cells remain alive despite cell cycle arrest.[80] Stem cells typically display tissue-specific differentiation patterns, and their ability to balance quiescence with proliferative activity is strongly regulated because this balance is critical for cell survival and the maintenance of regenerative responses.[158] Age-dependent reduction in the stem cell number or perturbed cell cycle activity has been reported in various stem cell types, including skeletal muscle, neural, and germ line stem cells.[163,164] Moreover, studies on aged mice demonstrated an overall decrease in cell cycle activity, including decreased cell divisions, in hematopoietic stem cells, together with an increased amount of DNA damage and telomere shortening, and an increased expression of cell cycle inhibitor p16, which also relates to senescence.[5,165] However, not only can cell cycle arrest and senescence lead to stem cell loss, increased stem cell proliferation and/or loss of quiescence of the stem cells can result in accelerated stem cell exhaustion and aging (Fig. 6.3).

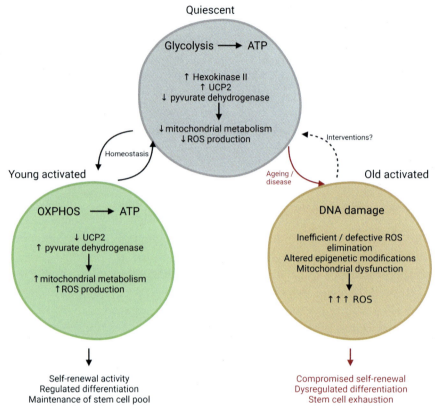

Figure 6.3 *Changes in cell metabolism in young and old stem cells. In stem cells residing in a quiescent state, glycolysis is the main pathway generating ATP. Quiescent cells express increased levels of hexokinase II and uncoupling protein 2 (UCP2), which blocks pyruvate metabolism in the mitochondria, leading to decreased mitochondrial metabolism and decreased reactive oxygen species (ROS) production. In the young activated stem cells, oxidative phosphorylation (OXPHOS) is the main generator of ATP, so mitochondrial metabolism is increased, and therefore ROS production also increases. This homeostasis between quiescence and activation seen in the young cells ensures self-renewal activity, tight regulation of differentiation and maintenance of the stem cell pool. However, in aging, this homeostasis between quiescence and activation is compromised, which leads to increased DNA damage owing to inefficient ROS elimination. Moreover, DNA damage results in altered epigenetic modifications and mitochondrial dysfunction, which creates a vicious cycle, creating even more ROS. Finally, it culminates in dysfunctional self-renewal activity, dysregulated differentiation, and finally stem cell exhaustion. Interventional strategies that can circumvent these age-induced defects are of great interest, among others, for mesenchymal stem cell–based therapies.*

A key regulator of the stem cell state (quiescence, proliferating, and differentiating) is cellular metabolism and the level of ROS (Fig. 6.3). In pluripotent stem cells (PSCs), ATP is produced mainly by glycolysis or oxidative phosphorylation (OXPHOS), depending on the state of the cell. During pluripotency and quiescence, glycolysis is the main pathway producing ATP, owing to a high expression level of hexokinase II and uncoupling protein 2 (UCP2) and decreased activity of pyruvate dehydrogenase, which converts pyruvate into mitochondrial acetyl coenzyme A. UCP2 prevents pyruvate from mitochondrial processing and blocks mitochondrial ATP production, although the exact mechanisms for pluripotency-related glycolysis are not fully understood.[166] When PSCs shift from pluripotency to differentiation, the cellular metabolism shifts to OXPHOS-mediated ATP production, which is associated with a decline in UCP2 expression. Furthermore, several studies demonstrated the ROS level to be an essential trigger of differentiation. Decreased levels of ROS inhibit differentiation, whereas increased levels of ROS induce differentiation. During aging, the regulation and function of the mitochondria, which are a main source of ROS, decline. Accumulation of damaged mitochondria are associated with an extensive increase and inefficient removal of ROS. This can result in increased damage to both nuclear and mitochondrial DNA, as well as to other macromolecules. Left unrepaired, this damage will lead to cellular dysfunctions, which will promote stem cell exhaustion and tissue degeneration, resulting in stem cell and organismal aging.[167]

Extensive research in the area of stem cell function and the discovery of the mechanisms controlling the pluripotency and generation of induced PSCs have provided ideas for therapies preventing stem cell exhaustion. However, a breakthrough is still missing in ways to correct the stem cell exhaustion phenotype, likely because of the complex, multifactorial nature of aging stem cells.[168]

4.1 Stem cell therapy and senescence

Mesenchymal stem cells (MSCs) are a line of stem cells found in multiple tissues of the body including the bone marrow and fat tissues, which can differentiate into cells from both the ectodermal and endodermal parentage.[169] MSCs holds therapeutic potential owing to their multipotency, their low immunogenicity, which means that allogeneic MSCs can be safely transplanted with a low risk for rejection by the recipient, relatively easy isolation from tissues, and their ability to grow in culture in vitro, which is needed for expansion of the cells.[170] However, the age of donors has a tremendous impact on the performance of MSCs in vitro. MSCs from aged individuals show decreased proliferation potential and rate, and turn senescent after fewer cell divisions than do cells from younger donors.[170] Stem cells can undergo symmetric cell division, in which one cell divides into two identical cells, both of which are capable of self-renewal, or they can divide asymmetrically. In asymmetric cell division, one cell divides into one stem cell with self-renewal capacity and one cell to be differentiated. In MSCs in culture, asymmetric cell division will result in one stem cell with proliferation capacity and one cell without it, which will turn senescent. Asymmetric cell division of MSCs in culture will therefore result in an accumulation of senescent cells with increased cell divisions. This is thought to be due to both epigenetic changes, such as methylation status of the DNA and histones, and to telomere attrition, as explained in detail earlier in this chapter. However, the detailed mechanisms and how cellular metabolism interferes with the senescent state of MSCs remain obscure. An increased proportion of senescent stem cells, including MSCs, results in decreased self-renewal and differentiation potential of the stem cell pool, which is also associated with aging.[170,171] In line with this, aged MSCs developed an SASP including proinflammatory markers such as IL-6, IL-8, and MCP1.[170,172] It is therefore necessary to understand the mechanisms of senescent development in MSCs and the markers of senescent MSCs to use the potential of MSC-based therapy fully and remove senescent cells or inhibit senescence development.

5. To develop senotherapies (senolytics and senomorphics) against age-related diseases and beyond

A type of agent that specifically eliminates senescent cells is termed a "senolytic": combining *seno-* from the word "senescence" with *-lytic,* a Greek term meaning "destroying." Senolytic treatments exploit the dependence of senescent cells on specific prosurvival pathways. Thus, senolytics transiently disable the prosurvival networks that protect senescent cells against their own apoptotic environment without affecting proliferating, quiescent, or differentiated cells.[173] Some senolytics have been tested, including dasatinib (a US Food and Drug Administration—approved tyrosine kinase inhibitor), quercetin (a flavonoid present in many fruits and vegetables), navitoclax A1331852 and A1155463 (Bcl-2 prosurvival family inhibitors) and fistein (a flavonoid). Preclinical studies in mice have shown that senolytics can eliminate senescent cells, resulting in the delay, prevention, or alleviation of multiple age- and senescence-related conditions[174,175] and extending the health span and life span.[7,176] The first in-human clinical trial of the senolytic cocktail of dasatinib and quercetin in idiopathic pulmonary fibrosis, a fatal cellular senescence-associated disease, was published. Unfortunately, the study showed no conclusive evidence of senescence clearance or reduction in SASP.[177] Another type of agent that targets senescent cells is termed senomorphics. Senomorphics do not eliminate senescent cells; instead, they target the SASP or other features of senescence and reverse the senescence phenotype.[178] Some identified molecules acting as senomorphics are the inhibitors of IκB kinase and NF-κB, and even the mTOR inhibitor rapamycin.[178,179] More studies and trials are needed to understand the effects of both senolytic and senomorphic therapy in humans and to examine the use of sonotherapy as a way to obtain healthy aging.

6. Methods for analyzing and quantifying senescence in vitro and in vivo

The more senescence is understood, the more complex and heterogeneous its phenotype appears. This makes it more difficult to define cellular senescence universally, especially in complex tissues and living organisms. Despite these challenges, new biomarkers and detection methods for cellular senescence are developing. With the tools we have today, the best way to validate cellular senescence with greater confidence is through the simultaneous use of different markers (for a detailed workflow see Kohli et al.[180]).

6.1 Methods for evaluating senescence in vitro

(A) Morphologic changes: In cell culture, it is possible to observe morphologic changes in senescent cells. Senescent cells are often larger and have a flatter appearance than nonsenescent cells, which can be observed with bright-field microscopy.[181] These changes in morphology have been described not only in mammalian cells, such as mice and humans, but also in simpler organisms such as yeast.[182,183]

(B) Cell cycle arrest: As described earlier, an important hallmark of cellular senescence is cell cycle arrest.[184] Of several markers for evaluating the state of cell cycle arrest, the most common are p16, p21, and p53, which are generally increased in senescence, and the phosphorylation of Rb, which is normally decreased owing to increased p16, as described previously.[184,185] These markers can be measured using different techniques such as quantitative PCR (qPCR), western blot, and immunohistochemistry. However, cell cycle arrest is not exclusive to senescence, and changes in those markers may be linked to other mechanisms of cell cycle arrest.[186,187] Therefore, they should be evaluated in conjunction with cellular proliferation and/or DNA replication assays.

(C) Nuclear changes: Heterochromatin and associated proteins change their configuration in cellular senescence. In nonsenescent cells, chromatin is allocated in the nuclear periphery and is less dense than in senescent cells. Senescence is associated with an enrichment of heterochromatin-associated proteins, such as heterochromatin protein 1, trimethylation of lysine 9 on histone H3, and the histone H2A variant macroH2A.[188,189] In addition, the condense chromatin in senescent cells contain structures called senescence-associated heterochromatin foci (SAHFs).[189] SAHFs contribute to the continuous cell cycle arrest in many senescent cells by repressing the expression of proliferation-promoting genes such as cyclin A.[189] By staining with the DNA-specific stain DAPI, SAHFs can be visualized even though it is not a SAHF-specific dye. In proliferating cells, DAPI staining will give a bright stain of the entire DNA-containing nucleoplasm. However, in senescent cells, SAHFs appear as compact, punctate DAPI stains, which can be costained alongside other heterochromatin-associated proteins and imaged with a confocal or fluorescence microscope.[189] Another nucleus-related change is the loss of or decrease in nuclear lamin B1. This can be measured as a decrease in lamin B1 signal by immunohistochemistry and is also associated with a decrease in the circularity of the nucleus.[123] Finally, detection of telomere shortening using telomere specific primers in qPCR or telomere-specific antibodies using fluorescence in situ hybridisation (FISH) techniques can be used to evaluate senescent cells.[3,190]

(D) β-Galactosidase activity: Increased activity of lysosomal β-Gal has been widely used as a marker of senescence. However, as with other senescence markers, it is not exclusively a senescence marker. β-Gal activity is also high in secretory cell types, such as macrophages, osteoclasts, and some cancer cells.[180] In addition, pH must be carefully controlled because commercially available kits detect SA-β-Gal only at pH 6.0. Furthermore, senescent cells accumulate high levels of lipofuscin (LF) in their lysosomes. LFs are aggregated, autofluorescent pigments formed by lipids, metals, and misfolded proteins.[191] LF can be used as a marker of senescence and can be costained with proliferation markers such as Ki-67 or with EdU staining (senescent cells will show LF accumulation and decreased signal from the proliferation marker). To stain for LF, either Sudan Black-B staining or the specific GL13-reagent can be used.[180] GL13 is a so-called superior biotinylated SBB analogue that binds to LF. Thus, normal immunohistochemistry with antibodies targeting the biotin on GL13 can be used to visualize the GL13-LF complex in the tissue or cell of interest.[180]

(E) Secretory phenotype: It can be complicated to measure or quantify the SASP owing to the several factors involved. Therefore, tools such as the SASP atlas[192] can help decide which markers to choose for a particular experiment.[193] SASP can be measured using qPCR for transcripts of some SASP components such as p53, p16, or p21, enzyme-linked immunosorbent assay (ELISA) kits measuring ILs such as IL-1α, IL-1β, and IL-6, or cytokine/protein arrays targeting, for instance, the family of MMPs, as introduced earlier in the discussion on SASP and summarized in Table 6.1.[194]

(F) Mitochondrial function, metabolic changes, and others: Although mitochondrial changes are not commonly used as a tool to detect senescence, mitochondrial dysfunction has an important role in preserving proinflammatory senescent phenotypes. Parameters of mitochondrial function that can be interesting in relation to senescence include the number, size, and morphology of mitochondria.[136,195] An example of a mitochondrial targeting probe is cyanine dye (CyBC9), a small nontoxic fluorescent molecule that specifically labels senescent MSCs. Senescence of MSCs leads to decreased mitochondrial membrane potential. This allows CyBC9, a membrane-permeable molecule, to accumulate in the mitochondria and mark senescent MSCs.[196] An increase in ROS, considered another hallmark of senescence, can be triggered by structural changes in mitochondria.[197] Contrary to mitochondrial probes, numerous dyes can be used to detect mitochondria-derived ROS, many of which are compatible with common laboratory instruments such as flow cytometers and spectrophotometers.[184,197] Other key determinants of mitochondrial health and the senescence state are the levels of NAD^+ and the $NAD^+/NADH$ ratio.[198,199] Both NAD^+ and the $NAD^+/NADH$ ratio can be measured by commercial kits or mass spectrometry.

6.2 Evaluation of senescence in vivo

Despite the significant effort to develop strategies to detect and assess senescence, studying senescence in vivo with high levels of confidence is challenging. In contrast to cell culture, most cells in living animals are quiescent or differentiated.[200] Therefore, senescence markers used in vitro, such as a lack of proliferation or DNA synthesis, cannot be used in vivo. Moreover, the complexity of the interaction among different cells in a whole organism does not allow the identification of morphologic changes observed in senescent cultured cells.[184]

(A) Ex vivo: Despite these challenges and considerations, the International Cellular Senescence Association described a recommended workflow for detecting senescence in tissue, including the use of multiple markers when evaluating senescence.[201]

(B) Living animals: Techniques to detect senescence in live animals have greatly improved. Several senescence reporter mouse models have been generated that allow the real-time detection and analysis of senescent cells in living organisms. One of these models is the p16-3MR mouse, which can be used to detect senescence through its expression of a trimodal reporter constituted of functional domains for Renilla luciferase, monomeric red fluorescent protein, and truncated herpes simplex virus 1 thymidine kinase.[78] It is also possible to use one of the broad number of chromogenic and fluorescent molecules and tracers detecting changes in SA-β-Gal. Some of them, such as a naphthalimide-based two-photon probe and the near-infrared-O^6-benzylguanine fluorescent molecule, have been successfully used to detect senescence in vivo. Another approach to detecting SA-β-Gal in vivo is to use radioactive probes, which may have potential translatability to human settings. For example, the tracer FPyGal used in positron emission tomography scans showed a correlation between tracer uptake in chemotherapy-treated tumors and SA-β-Gal activity. These tools show promise for detecting the state of senescence in living organisms in the near future.[184]

7. Concluding remarks

Cellular senescence leads to cell growth arrest and is generally associated with a multifaceted SASP. Despite the heterogeneity of senescence, some of the most common characteristics of senescence include cell cycle arrest associated with increased p53, p21, and p16, increased cell and nuclear size, DNA and telomere damaged foci, and mitochondrial accumulation. These parameters and markers can be used to identify and quantify senescence both in vitro and in vivo. In aging, various stimuli can lead to senescence. Genomic instability and DNA damage increase with age and can lead to a persistent DDR, which can induce senescence. In line with this, telomere shortening and attrition as well as epigenetic modifications can lead to senescence. Also, the activation of tumor suppressors and the over-expression of certain oncogenes can result in senescence, demonstrating the wide range of senescence-inducing stimuli. Senescence is a hallmark of aging that has a role in several age-predisposed diseases including neuro-degenerative diseases and cancers. Moreover, the regulation and prolifera-tion of stem cells during aging declines, likely owing to increased senescence in the stem cell—containing tissues. Preclinical studies suggest that eliminating senescent cells hampers aging and at least some age-associated diseases. However, caution much be taken, because complete removal of senescent cells might stimulate cancer growth.

We are still at the tip of the iceberg when it comes to understanding the roles of senescence in aging, how SASP evolves, and what consequences SASP has under both paracrine and autocrine conditions. Furthermore, we need more studies on senescence to be able to distinguish what happens during senescence in postmitotic senescence, such as neurons versus proliferating cells, pathologic senescence, and physiologic senescence. This will help to develop senolytic/senomorphic-based therapy against aging-related phenomena and diseases.

Acknowledgments

The authors acknowledge the valuable work of the many investigators whose published articles they were unable to cite owing to space limitations. They thank Thale Dawn Patrick-Brown for critical reading of the manuscript. E.F.F. was supported by the National Natural Science Foundation of China (No. 81971327), HELSE SØR-ØST (Nos. 2017056, 2020001, and 2021021), the Research Council of Norway (No. 262175), Akershus University Hospital (Nos. 269901, 261973, and 262960), the Civitan Norges Forskningsfond for Alzheimer's Sykdom (No. 281931), the Czech Republic—Norway KAPPA program (with Martin Vyhnálek, No. TO01000215), and the Rosa Sløyfe/Norwegian Cancer Society and

Norwegian Breast Cancer Society (No. 207819). S.L. has received funding from the European Union's Horizon 2020 research and innovation program under the Marie Skłodowska-Curie Grant Agreement, No 801133.

Competing interests statement

E.F.F. has a CRADA arrangement with ChromaDex (United States) and is consultant to Aladdin Healthcare Technologies (United Kingdom and Germany), the Vancouver Dementia Prevention Center (Canada), Intellectual Labs (Norway), and MindRank AI (China).

References

1. Hayflick L. The limited in vitro lifetime of human diploid cell strains. *Exp Cell Res.* 1965;37:614−636. https://doi.org/10.1016/0014-4827(65)90211-9 ([published Online First: Epub Date]|).
2. Muñoz-Espín D, Cañamero M, Maraver A, et al. Programmed cell senescence during mammalian embryonic development. *Cell.* 2013;155(5):1104−1118. https://doi.org/10.1016/j.cell.2013.10.019 ([published Online First: Epub Date]|).
3. Di Micco R, Krizhanovsky V, Baker D, d'Adda di Fagagna F. Cellular senescence in ageing: from mechanisms to therapeutic opportunities. *Nat Rev Mol Cell Biol.* 2021;22(2):75−95. https://doi.org/10.1038/s41580-020-00314-w ([published Online First: Epub Date]|).
4. Coppé JP, Desprez PY, Krtolica A, Campisi J. The senescence-associated secretory phenotype: the dark side of tumor suppression. *Annu Rev Pathol.* 2010;5:99−118. https://doi.org/10.1146/annurev-pathol-121808-102144 ([published Online First: Epub Date]|).
5. López-Otín C, Blasco MA, Partridge L, Serrano M, Kroemer G. The hallmarks of aging. *Cell.* 2013;153(6):1194−1217. https://doi.org/10.1016/j.cell.2013.05.039 ([published Online First: Epub Date]|).
6. Rodier F, Campisi J. Four faces of cellular senescence. *J Cell Biol.* 2011;192(4):547−556. https://doi.org/10.1083/jcb.201009094 ([published Online First: Epub Date]|).
7. Xu M, Pirtskhalava T, Farr JN, et al. Senolytics improve physical function and increase lifespan in old age. *Nat Med.* 2018;24(8):1246−1256. https://doi.org/10.1038/s41591-018-0092-9 ([published Online First: Epub Date]|).
8. Iske J, Seyda M, Heinbokel T, et al. Senolytics prevent mt-DNA-induced inflammation and promote the survival of aged organs following transplantation. *Nat Commun.* 2020;11(1):4289. https://doi.org/10.1038/s41467-020-18039-x ([published Online First: Epub Date]|).
9. Rodier F, Coppé JP, Patil CK, et al. Persistent DNA damage signalling triggers senescence-associated inflammatory cytokine secretion. *Nat Cell Biol.* 2009;11(8):973−979. https://doi.org/10.1038/ncb1909 ([published Online First: Epub Date]|).
10. Fang EF, Scheibye-Knudsen M, Chua KF, Mattson MP, Croteau DL, Bohr VA. Nuclear DNA damage signalling to mitochondria in ageing. *Nat Rev Mol Cell Biol.* 2016;17(5):308−321. https://doi.org/10.1038/nrm.2016.14 ([published Online First: Epub Date]|).
11. Hernandez-Segura A, Nehme J, Demaria M. Hallmarks of cellular senescence. *Trend Cell Biol.* 2018;28(6):436−453. https://doi.org/10.1016/j.tcb.2018.02.001 ([published Online First: Epub Date]|).

12. Dimri GP, Hara E, Campisi J. Regulation of two E2F-related genes in presenescent and senescent human fibroblasts. *J Biol Chem*. 1994;269(23):16180–16186.
13. Campisi J. Aging and cancer: the double-edged sword of replicative senescence. *J Am Geriatr Soc*. 1997;45(4):482–488. https://doi.org/10.1111/j.1532-5415.1997.tb051 75.x ([published Online First: Epub Date]|).
14. Coppé JP, Patil CK, Rodier F, et al. A human-like senescence-associated secretory phenotype is conserved in mouse cells dependent on physiological oxygen. *PloS One*. 2010;5(2):e9188. https://doi.org/10.1371/journal.pone.0009188 ([published Online First: Epub Date]|).
15. Campisi J. Aging, cellular senescence, and cancer. *Annu Rev Physiol*. 2013;75:685–705. https://doi.org/10.1146/annurev-physiol-030212-183653 ([published Online First: Epub Date]|).
16. Rufini A, Tucci P, Celardo I, Melino G. Senescence and aging: the critical roles of p53. *Oncogene*. 2013;32(43):5129–5143. https://doi.org/10.1038/onc.2012.640 ([published Online First: Epub Date]|).
17. Maier B, Gluba W, Bernier B, et al. Modulation of mammalian life span by the short isoform of p53. *Genes Dev*. 2004;18(3):306–319. https://doi.org/10.1101/gad.1162404 ([published Online First: Epub Date]|).
18. Marcel V, Dichtel-Danjoy ML, Sagne C, et al. Biological functions of p53 isoforms through evolution: lessons from animal and cellular models. *Cell Death Differ*. 2011;18(12):1815–1824. https://doi.org/10.1038/cdd.2011.120 ([published Online First: Epub Date]|).
19. Kenyon CJ. The genetics of ageing. *Nature*. 2010;464(7288):504–512. https://doi.org/10.1038/nature08980 ([published Online First: Epub Date]|).
20. Bischoff FZ, Yim SO, Pathak S, et al. Spontaneous abnormalities in normal fibroblasts from patients with Li-Fraumeni cancer syndrome: aneuploidy and immortalization. *Cancer Res*. 1990;50(24):7979–7984.
21. Harvey M, McArthur MJ, Montgomery Jr CA, Butel JS, Bradley A, Donehower LA. Spontaneous and carcinogen-induced tumorigenesis in p53-deficient mice. *Nat Genet*. 1993;5(3):225–229. https://doi.org/10.1038/ng1193-225 ([published Online First: Epub Date]|).
22. Shay JW, Pereira-Smith OM, Wright WE. A role for both RB and p53 in the regulation of human cellular senescence. *Exp Cell Res*. 1991;196(1):33–39. https://doi.org/10.1016/0014-4827(91)90453-2 ([published Online First: Epub Date]|).
23. Harley CB, Kim NW, Prowse KR, et al. Telomerase, cell immortality, and cancer. *Cold Spring Harb Symp Quant Biol*. 1994;59:307–315. https://doi.org/10.1101/sqb.1994.059.01.035 ([published Online First: Epub Date]|).
24. Talens F, Van Vugt M. Inflammatory signaling in genomically instable cancers. *Cell Cycle*. 2019;18(16):1830–1848. https://doi.org/10.1080/15384101.2019.1638192 ([published Online First: Epub Date]|).
25. Mijit M, Caracciolo V, Melillo A, Amicarelli F, Giordano A. Role of p53 in the regulation of cellular senescence. *Biomolecules*. 2020;10(3):420. https://doi.org/10.3390/biom10030420 ([published Online First: Epub Date]|).
26. Chang BD, Watanabe K, Broude EV, et al. Effects of p21Waf1/Cip1/Sdi1 on cellular gene expression: implications for carcinogenesis, senescence, and age-related diseases. *Proc Nat Acad Sci U S A*. 2000;97(8):4291–4296. https://doi.org/10.1073/pnas.97.8.4291 ([published Online First: Epub Date]|).
27. McConnell BB, Starborg M, Brookes S, Peters G. Inhibitors of cyclin-dependent kinases induce features of replicative senescence in early passage human diploid fibroblasts. *Curr Biol*. 1998;8(6):351–354. https://doi.org/10.1016/s0960-9822(98)70137-x ([published Online First: Epub Date]|).

28. Brown JP, Wei W, Sedivy JM. Bypass of senescence after disruption of p21CIP1/WAF1 gene in normal diploid human fibroblasts. *Science*. 1997;277(5327):831–834. https://doi.org/10.1126/science.277.5327.831 ([published Online First: Epub Date]|).

29. Pantoja C, Serrano M. Murine fibroblasts lacking p21 undergo senescence and are resistant to transformation by oncogenic Ras. *Oncogene*. 1999;18(35):4974–4982. https://doi.org/10.1038/sj.onc.1202880 ([published Online First: Epub Date]|).

30. Groth A, Weber JD, Willumsen BM, Sherr CJ, Roussel MF. Oncogenic Ras induces p19ARF and growth arrest in mouse embryo fibroblasts lacking p21Cip1 and p27Kip1 without activating cyclin D-dependent kinases. *J Biol Chem*. 2000;275(35):27473–27480. https://doi.org/10.1074/jbc.M003417200 ([published Online First: Epub Date]|).

31. Wei W, Hemmer RM, Sedivy JM. Role of p14(ARF) in replicative and induced senescence of human fibroblasts. *Mol Cell Biol*. 2001;21(20):6748–6757. https://doi.org/10.1128/mcb.21.20.6748-6757.2001 ([published Online First: Epub Date]|).

32. Beauséjour CM, Krtolica A, Galimi F, et al. Reversal of human cellular senescence: roles of the p53 and p16 pathways. *Embo J*. 2003;22(16):4212–4222. https://doi.org/10.1093/emboj/cdg417 (published Online First: Epub Date]|).

33. Krishnamurthy J, Torrice C, Ramsey MR, et al. Ink4a/Arf expression is a biomarker of aging. *J Clin Invest*. 2004;114(9):1299–1307. https://doi.org/10.1172/jci22475 ([published Online First: Epub Date]|).

34. Ressler S, Bartkova J, Niederegger H, et al. p16INK4A is a robust in vivo biomarker of cellular aging in human skin. *Aging Cell*. 2006;5(5):379–389. https://doi.org/10.1111/j.1474-9726.2006.00231.x ([published Online First: Epub Date]|).

35. Jeck WR, Siebold AP, Sharpless NE. Review: a meta-analysis of GWAS and age-associated diseases. *Aging Cell*. 2012;11(5):727–731. https://doi.org/10.1111/j.1474-9726.2012.00871.x ([published Online First: Epub Date]|).

36. Sharpless NE, Ramsey MR, Balasubramanian P, Castrillon DH, DePinho RA. The differential impact of p16(INK4a) or p19(ARF) deficiency on cell growth and tumorigenesis. *Oncogene*. 2004;23(2):379–385. https://doi.org/10.1038/sj.onc.1207074 ([published Online First: Epub Date]|).

37. Coppé JP, Patil CK, Rodier F, et al. Senescence-associated secretory phenotypes reveal cell-nonautonomous functions of oncogenic RAS and the p53 tumor suppressor. *PLoS Biol*. 2008;6(12):2853–2868. https://doi.org/10.1371/journal.pbio.0060301 ([published Online First: Epub Date]|).

38. Funayama R, Ishikawa F. Cellular senescence and chromatin structure. *Chromosoma*. 2007;116(5):431–440. https://doi.org/10.1007/s00412-007-0115-7 ([published Online First: Epub Date]|).

39. Mehta IS, Figgitt M, Clements CS, Kill IR, Bridger JM. Alterations to nuclear architecture and genome behavior in senescent cells. *Ann N Y Acad Sci*. 2007;1100:250–263. https://doi.org/10.1196/annals.1395.027 ([published Online First: Epub Date]|).

40. Narita M, Nũnez S, Heard E, et al. Rb-mediated heterochromatin formation and silencing of E2F target genes during cellular senescence. *Cell*. 2003;113(6):703–716. https://doi.org/10.1016/s0092-8674(03)00401-x ([published Online First: Epub Date]|).

41. Kuilman T, Michaloglou C, Vredeveld LC, et al. Oncogene-induced senescence relayed by an interleukin-dependent inflammatory network. *Cell*. 2008;133(6):1019–1031. https://doi.org/10.1016/j.cell.2008.03.039 ([published Online First: Epub Date]|).

42. Lu SY, Chang KW, Liu CJ, et al. Ripe areca nut extract induces G1 phase arrests and senescence-associated phenotypes in normal human oral keratinocyte. *Carcinogenesis*.

2006;27(6):1273–1284. https://doi.org/10.1093/carcin/bgi357 ([published Online First: Epub Date]|).

43. Sarkar D, Lebedeva IV, Emdad L, Kang DC, Baldwin Jr AS, Fisher PB. Human polynucleotide phosphorylase (hPNPaseold-35): a potential link between aging and inflammation. *Canc Res.* 2004;64(20):7473–7478. https://doi.org/10.1158/0008-5472.Can-04-1772 ([published Online First: Epub Date]|).

44. Bhaumik D, Scott GK, Schokrpur S, et al. MicroRNAs miR-146a/b negatively modulate the senescence-associated inflammatory mediators IL-6 and IL-8. *Aging (Albany NY).* 2009;1(4):402–411. https://doi.org/10.18632/aging.100042 ([published Online First: Epub Date]|).

45. Orjalo AV, Bhaumik D, Gengler BK, Scott GK, Campisi J. Cell surface-bound IL-1alpha is an upstream regulator of the senescence-associated IL-6/IL-8 cytokine network. *Proc Na Acad Sci U S A.* 2009;106(40):17031–17036. https://doi.org/10.1073/pnas.0905299106 ([published Online First: Epub Date]|).

46. Maier JA, Voulalas P, Roeder D, Maciag T. Extension of the life-span of human endothelial cells by an interleukin-1 alpha antisense oligomer. *Science.* 1990;249(4976):1570–1574. https://doi.org/10.1126/science.2218499 ([published Online First: Epub Date]|).

47. Mantovani A, Locati M, Vecchi A, Sozzani S, Allavena P. Decoy receptors: a strategy to regulate inflammatory cytokines and chemokines. *Trends Immunol.* 2001;22(6):328–336. https://doi.org/10.1016/s1471-4906(01)01941-x ([published Online First: Epub Date]|).

48. Huang NN, Wang DJ, Heppel LA. Stimulation of aged human lung fibroblasts by extracellular ATP via suppression of arachidonate metabolism. *J Biol Chem.* 1993;268(15):10789–10795.

49. Wang S, Moerman EJ, Jones RA, Thweatt R, Goldstein S. Characterization of IGFBP-3, PAI-1 and SPARC mRNA expression in senescent fibroblasts. *Mech Ageing Dev.* 1996;92(2–3):121–132. https://doi.org/10.1016/s0047-6374(96)01814-3 ([published Online First: Epub Date]|).

50. Grillari J, Hohenwarter O, Grabherr RM, Katinger H. Subtractive hybridization of mRNA from early passage and senescent endothelial cells. *Exp Gerontol.* 2000;35(2):187–197. https://doi.org/10.1016/s0531-5565(00)00080-2 ([published Online First: Epub Date]|).

51. Wajapeyee N, Serra RW, Zhu X, Mahalingam M, Green MR. Oncogenic BRAF induces senescence and apoptosis through pathways mediated by the secreted protein IGFBP7. *Cell.* 2008;132(3):363–374. https://doi.org/10.1016/j.cell.2007.12.032 ([published Online First: Epub Date]|).

52. Hornebeck W, Maquart FX. Proteolyzed matrix as a template for the regulation of tumor progression. *Biomed Pharmacother.* 2003;57(5–6):223–230. https://doi.org/10.1016/s0753-3322(03)00049-0 ([published Online First: Epub Date]|).

53. McQuibban GA, Gong JH, Wong JP, Wallace JL, Clark-Lewis I, Overall CM. Matrix metalloproteinase processing of monocyte chemoattractant proteins generates CC chemokine receptor antagonists with anti-inflammatory properties in vivo. *Blood.* 2002;100(4):1160–1167.

54. Van Den Steen PE, Wuyts A, Husson SJ, Proost P, Van Damme J, Opdenakker G. Gelatinase B/MMP-9 and neutrophil collagenase/MMP-8 process the chemokines human GCP-2/CXCL6, ENA-78/CXCL5 and mouse GCP-2/LIX and modulate their physiological activities. *Eur J Biochem.* 2003;270(18):3739–3749. https://doi.org/10.1046/j.1432-1033.2003.03760.x ([published Online First: Epub Date]|).

55. Blasi F, Carmeliet P. uPAR: a versatile signalling orchestrator. *Nat Rev Mol Cell Biol.* 2002;3(12):932–943. https://doi.org/10.1038/nrm977 ([published Online First: Epub Date]|).

56. Kortlever RM, Higgins PJ, Bernards R. Plasminogen activator inhibitor-1 is a critical downstream target of p53 in the induction of replicative senescence. *Nat Cell Biol.* 2006;8(8):877—884. https://doi.org/10.1038/ncb1448 ([published Online First: Epub Date]|).

57. Antia M, Baneyx G, Kubow KE, Vogel V. Fibronectin in aging extracellular matrix fibrils is progressively unfolded by cells and elicits an enhanced rigidity response. *Faraday Discuss.* 2008;139:229—249. https://doi.org/10.1039/b718714a ([published Online First: Epub Date]|).

58. Rasoamanantena P, Thweatt R, Labat-Robert J, Goldstein S. Altered regulation of fibronectin gene expression in Werner syndrome fibroblasts. *Exp Cell Res.* 1994;213(1):121—127. https://doi.org/10.1006/excr.1994.1181 (published Online First: Epub Date]|).

59. Kumazaki T, Kobayashi M, Mitsui Y. Enhanced expression of fibronectin during in vivo cellular aging of human vascular endothelial cells and skin fibroblasts. *Exp Cell Res.* 1993;205(2):396—402. https://doi.org/10.1006/excr.1993.1103 ([published Online First: Epub Date]|).

60. Lee AC, Fenster BE, Ito H, et al. Ras proteins induce senescence by altering the intracellular levels of reactive oxygen species. *J Biol Chem.* 1999;274(12):7936—7940. https://doi.org/10.1074/jbc.274.12.7936 ([published Online First: Epub Date]|).

61. Xin MG, Zhang J, Block ER, Patel JM. Senescence-enhanced oxidative stress is associated with deficiency of mitochondrial cytochrome c oxidase in vascular endothelial cells. *Mech Ageing Dev.* 2003;124(8—9):911—919. https://doi.org/10.1016/s0047-6374(03)00163-5 ([published Online First: Epub Date]|).

62. Finkel T, Holbrook NJ. Oxidants, oxidative stress and the biology of ageing. *Nature.* 2000;408(6809):239—247. https://doi.org/10.1038/35041687 ([published Online First: Epub Date]|).

63. Donehower LA. Does p53 affect organismal aging? *J Cell Physiol.* 2002;192(1):23—33. https://doi.org/10.1002/jcp.10104 ([published Online First: Epub Date]|).

64. Goldstein S. Replicative senescence: the human fibroblast comes of age. *Science.* 1990;249(4973):1129—1133. https://doi.org/10.1126/science.2204114 ([published Online First: Epub Date]|).

65. Martin GM, Sprague CA, Epstein CJ. Replicative life-span of cultivated human cells. Effects of donor's age, tissue, and genotype. *Lab Invest J Tech Meth Pathol.* 1970;23(1):86—92.

66. Norwood TH, Hoehn H, Salk D, Martin GM. Cellular aging in Werner's syndrome: a unique phenotype? *J Invest Dermatol.* 1979;73(1):92—96. https://doi.org/10.1111/1523-1747.ep12532778 ([published Online First: Epub Date]|).

67. Shay JW, Wright WE, Werbin H. Defining the molecular mechanisms of human cell immortalization. *Biochimica et biophysica acta.* 1991;1072(1):1—7. https://doi.org/10.1016/0304-419x(91)90003-4 ([published Online First: Epub Date]|).

68. Sager R, Tanaka K, Lau CC, Ebina Y, Anisowicz A. Resistance of human cells to tumorigenesis induced by cloned transforming genes. *Proc Nat Acad Sci U S A.* 1983;80(24):7601—7605. https://doi.org/10.1073/pnas.80.24.7601 ([published Online First: Epub Date]|).

69. Yousefzadeh MJ, Melos KI, Angelini L, Burd CE, Robbins PD, Niedernhofer LJ. Mouse models of accelerated cellular senescence. *Meth Mol Biol.* 2019;1896:203—230. https://doi.org/10.1007/978-1-4939-8931-7_17 ([published Online First: Epub Date]|).

70. Baker DJ, Childs BG, Durik M, et al. Naturally occurring p16(Ink4a)-positive cells shorten healthy lifespan. *Nature.* 2016;530(7589):184—189. https://doi.org/10.1038/nature16932 ([published Online First: Epub Date]|).

71. Baker DJ, Wijshake T, Tchkonia T, et al. Clearance of p16Ink4a-positive senescent cells delays ageing-associated disorders. *Nature.* 2011;479(7372):232—236. https://doi.org/10.1038/nature10600 ([published Online First: Epub Date]|).

72. Musi N, Valentine JM, Sickora KR, et al. Tau protein aggregation is associated with cellular senescence in the brain. *Aging Cell.* 2018;17(6):e12840. https://doi.org/10.1111/acel.12840 ([published Online First: Epub Date]|).

73. Santacruz K, Lewis J, Spires T, et al. Tau suppression in a neurodegenerative mouse model improves memory function. *Science.* 2005;309(5733):476—481. https://doi.org/10.1126/science.1113694 ([published Online First: Epub Date]|).

74. Bussian TJ, Aziz A, Meyer CF, Swenson BL, van Deursen JM, Baker DJ. Clearance of senescent glial cells prevents tau-dependent pathology and cognitive decline. *Nature.* 2018;562(7728):578—582. https://doi.org/10.1038/s41586-018-0543-y ([published Online First: Epub Date]|).

75. Demaria M, Ohtani N, Youssef SA, et al. An essential role for senescent cells in optimal wound healing through secretion of PDGF-AA. *Dev Cell.* 2014;31(6):722—733. https://doi.org/10.1016/j.devcel.2014.11.012 ([published Online First: Epub Date]|).

76. Baker DJ, Jeganathan KB, Cameron JD, et al. BubR1 insufficiency causes early onset of aging-associated phenotypes and infertility in mice. *Nat Genet.* 2004;36(7):744—749. https://doi.org/10.1038/ng1382 ([published Online First: Epub Date]|).

77. Baker DJ, Jin F, van Deursen JM. The yin and yang of the Cdkn2a locus in senescence and aging. *Cell Cycle.* 2008;7(18):2795—2802. https://doi.org/10.4161/cc.7.18.6687 ([published Online First: Epub Date]|).

78. Laberge RM, Adler D, DeMaria M, et al. Mitochondrial DNA damage induces apoptosis in senescent cells. *Cell Death Dis.* 2013;4(7):e727. https://doi.org/10.1038/cddis.2013.199 ([published Online First: Epub Date]|).

79. Lee S, Lee JS. Cellular senescence: a promising strategy for cancer therapy. *BMB Rep.* 2019;52(1):35—41. https://doi.org/10.5483/BMBRep.2019.52.1.294 ([published Online First: Epub Date]|).

80. Campisi J. Cancer, aging and cellular senescence. *In Vivo.* 2000;14(1):183—188.

81. Minamino T, Miyauchi H, Yoshida T, Ishida Y, Yoshida H, Komuro I. Endothelial cell senescence in human atherosclerosis: role of telomere in endothelial dysfunction. *Circulation.* 2002;105(13):1541—1544. https://doi.org/10.1161/01.cir.0000013836.85741.17 ([published Online First: Epub Date]|).

82. Roos CM, Zhang B, Palmer AK, et al. Chronic senolytic treatment alleviates established vasomotor dysfunction in aged or atherosclerotic mice. *Aging Cell.* 2016;15(5):973—977. https://doi.org/10.1111/acel.12458 ([published Online First: Epub Date]|).

83. Schafer MJ, White TA, Iijima K, et al. Cellular senescence mediates fibrotic pulmonary disease. *Nat Commun.* 2017;8:14532. https://doi.org/10.1038/ncomms14532 ([published Online First: Epub Date]|).

84. Ogrodnik M, Miwa S, Tchkonia T, et al. Cellular senescence drives age-dependent hepatic steatosis. *Nat Commun.* 2017;8:15691. https://doi.org/10.1038/ncomms15691 ([published Online First: Epub Date]|).

85. Baker DJ, Petersen RC. Cellular senescence in brain aging and neurodegenerative diseases: evidence and perspectives. *J Clin Invest.* 2018;128(4):1208—1216. https://doi.org/10.1172/jci95145 ([published Online First: Epub Date]|).

86. Siracusa R, Fusco R, Cuzzocrea S. Astrocytes: role and functions in brain pathologies. *Front Pharmacol.* 2019;10(1114). https://doi.org/10.3389/fphar.2019.01114 ([published Online First: Epub Date]|).

87. Organization WH. *Global Status Report on the Public Health Response to Dementia*; 2021. www.who.int/publications/i/item/9789240033245.
88. Kerr JS, Adriaanse BA, Greig NH, et al. Mitophagy and Alzheimer's disease: cellular and molecular mechanisms. *Trend Neurosci.* 2017;40(3):151–166. https://doi.org/10.1016/j.tins.2017.01.002 ([published Online First: Epub Date]|).
89. Fang EF, Hou Y, Palikaras K, et al. Mitophagy inhibits amyloid-beta and tau pathology and reverses cognitive deficits in models of Alzheimer's disease. *Nat Neurosci.* 2019;22(3):401–412. https://doi.org/10.1038/s41593-018-0332-9 ([published Online First: Epub Date]|).
90. Zhang P, Kishimoto Y, Grammatikakis I, et al. Senolytic therapy alleviates Aβ–associated oligodendrocyte progenitor cell senescence and cognitive deficits in an Alzheimer's disease model. *Nat Neurosci.* 2019;22(5):719–728. https://doi.org/10.1038/s41593-019-0372-9 ([published Online First: Epub Date]|).
91. Bhat R, Crowe EP, Bitto A, et al. Astrocyte senescence as a component of Alzheimer's disease. *PloS One.* 2012;7(9):e45069. https://doi.org/10.1371/journal.pone.0045069 (published Online First: Epub Date]|).
92. Flanary BE, Sammons NW, Nguyen C, Walker D, Streit WJ. Evidence that aging and amyloid promote microglial cell senescence. *Rejuvenation Res.* 2007;10(1):61–74. https://doi.org/10.1089/rej.2006.9096 ([published Online First: Epub Date]|).
93. Myung NH, Zhu X, Kruman II , et al. Evidence of DNA damage in Alzheimer disease: phosphorylation of histone H2AX in astrocytes. *Age (Dordr).* 2008;30(4):209–215. https://doi.org/10.1007/s11357-008-9050-7 ([published Online First: Epub Date]|).
94. Storer M, Mas A, Robert-Moreno A, et al. Senescence is a developmental mechanism that contributes to embryonic growth and patterning. *Cell.* 2013;155(5):1119–1130. https://doi.org/10.1016/j.cell.2013.10.041 ([published Online First: Epub Date]|).
95. Jun JI, Lau LF. The matricellular protein CCN1 induces fibroblast senescence and restricts fibrosis in cutaneous wound healing. *Nat Cell Biol.* 2010;12(7):676–685. https://doi.org/10.1038/ncb2070 ([published Online First: Epub Date]|).
96. Krizhanovsky V, Yon M, Dickins RA, et al. Senescence of activated stellate cells limits liver fibrosis. *Cell.* 2008;134(4):657–667. https://doi.org/10.1016/j.cell.2008.06.049 ([published Online First: Epub Date]|).
97. van Deursen JM. The role of senescent cells in ageing. *Nature.* 2014;509(7501):439–446. https://doi.org/10.1038/nature13193 ([published Online First: Epub Date]|).
98. Roninson IB. Tumor cell senescence in cancer treatment. *Cancer Res.* 2003;63(11):2705–2715.
99. Acosta JC, Banito A, Wuestefeld T, et al. A complex secretory program orchestrated by the inflammasome controls paracrine senescence. *Nat Cell Biol.* 2013;15(8):978–990. https://doi.org/10.1038/ncb2784 ([published Online First: Epub Date]|).
100. Jiang C, Liu G, Luckhardt T, et al. Serpine 1 induces alveolar type II cell senescence through activating p53-p21-Rb pathway in fibrotic lung disease. *Aging Cell.* 2017;16(5):1114–1124. https://doi.org/10.1111/acel.12643 ([published Online First: Epub Date]|).
101. Dulić V, Beney GE, Frebourg G, Drullinger LF, Stein GH. Uncoupling between phenotypic senescence and cell cycle arrest in aging p21-deficient fibroblasts. *Mol Cell Biol.* 2000;20(18):6741–6754. https://doi.org/10.1128/mcb.20.18.6741-6754.2000 ([published Online First: Epub Date]|).
102. d'Adda di Fagagna F, Reaper PM, Clay-Farrace L, et al. A DNA damage checkpoint response in telomere-initiated senescence. *Nature.* 2003;426(6963):194–198. https://doi.org/10.1038/nature02118 ([published Online First: Epub Date]|).

103. Herbig U, Jobling WA, Chen BP, Chen DJ, Sedivy JM. Telomere shortening triggers senescence of human cells through a pathway involving ATM, p53, and p21(CIP1), but not p16(INK4a). *Mol Cell.* 2004;14(4):501–513. https://doi.org/10.1016/s1097-2765(04)00256-4 ([published Online First: Epub Date]|).

104. Hemann MT, Strong MA, Hao LY, Greider CW. The shortest telomere, not average telomere length, is critical for cell viability and chromosome stability. *Cell.* 2001;107(1):67–77. https://doi.org/10.1016/s0092-8674(01)00504-9 ([published Online First: Epub Date]|).

105. Bodnar AG, Ouellette M, Frolkis M, et al. Extension of life-span by introduction of telomerase into normal human cells. *Science.* 1998;279(5349):349–352. https://doi.org/10.1126/science.279.5349.349 ([published Online First: Epub Date]|).

106. d'Adda di Fagagna F. Living on a break: cellular senescence as a DNA-damage response. *Nat Rev Cancer.* 2008;8(7):512–522. https://doi.org/10.1038/nrc2440 ([published Online First: Epub Date]|).

107. de Lange T. Shelterin-mediated telomere protection. *Annu Rev Genet.* 2018;52:223–247. https://doi.org/10.1146/annurev-genet-032918-021921 ([published Online First: Epub Date]|).

108. Fumagalli M, Rossiello F, Clerici M, et al. Telomeric DNA damage is irreparable and causes persistent DNA-damage-response activation. *Nat Cell Biol.* 2012;14(4):355–365. https://doi.org/10.1038/ncb2466 ([published Online First: Epub Date]|).

109. Hewitt G, Jurk D, Marques FD, et al. Telomeres are favoured targets of a persistent DNA damage response in ageing and stress-induced senescence. *Nat Communicat.* 2012;3:708. https://doi.org/10.1038/ncomms1708 (published Online First: Epub Date]|).

110. Bae NS, Baumann P. A RAP1/TRF2 complex inhibits nonhomologous end-joining at human telomeric DNA ends. *Mol Cell.* 2007;26(3):323–334. https://doi.org/10.1016/j.molcel.2007.03.023 ([published Online First: Epub Date]|).

111. Rossiello F, Aguado J, Sepe S, et al. DNA damage response inhibition at dysfunctional telomeres by modulation of telomeric DNA damage response RNAs. *Nat Commun.* 2017;8:13980. https://doi.org/10.1038/ncomms13980 ([published Online First: Epub Date]|).

112. Herbst A, Pak JW, McKenzie D, Bua E, Bassiouni M, Aiken JM. Accumulation of mitochondrial DNA deletion mutations in aged muscle fibers: evidence for a causal role in muscle fiber loss. *J Gerontol A Biol Sci Med Sci.* 2007;62(3):235–245. https://doi.org/10.1093/gerona/62.3.235 ([published Online First: Epub Date]|).

113. Lee HY, Choi CS, Birkenfeld AL, et al. Targeted expression of catalase to mitochondria prevents age-associated reductions in mitochondrial function and insulin resistance. *Cell Metabol.* 2010;12(6):668–674. https://doi.org/10.1016/j.cmet.2010.11.004 ([published Online First: Epub Date]|).

114. Safdar A, Hamadeh MJ, Kaczor JJ, Raha S, Debeer J, Tarnopolsky MA. Aberrant mitochondrial homeostasis in the skeletal muscle of sedentary older adults. *PloS One.* 2010;5(5):e10778. https://doi.org/10.1371/journal.pone.0010778 ([published Online First: Epub Date]|).

115. Braig M, Schmitt CA. Oncogene-induced senescence: putting the brakes on tumor development. *Cancer Res.* 2006;66(6):2881–2884. https://doi.org/10.1158/0008-5472.Can-05-4006 ([published Online First: Epub Date]|).

116. Campisi J, d'Adda di Fagagna F. Cellular senescence: when bad things happen to good cells. *Nat Rev Mol Cell Biol.* 2007;8(9):729–740. https://doi.org/10.1038/nrm2233 ([published Online First: Epub Date]|).

117. Passos JF, Nelson G, Wang C, et al. Feedback between p21 and reactive oxygen production is necessary for cell senescence. *Mol Syst Biol.* 2010;6:347. https://doi.org/10.1038/msb.2010.5 ([published Online First: Epub Date]|).

118. Kaplon J, Zheng L, Meissl K, et al. A key role for mitochondrial gatekeeper pyruvate dehydrogenase in oncogene-induced senescence. *Nature.* 2013;498(7452):109−112. https://doi.org/10.1038/nature12154 ([published Online First: Epub Date]|).

119. Moiseeva O, Bourdeau V, Roux A, Deschênes-Simard X, Ferbeyre G. Mitochondrial dysfunction contributes to oncogene-induced senescence. *Mol Cell Biol.* 2009;29(16):4495−4507. https://doi.org/10.1128/mcb.01868-08 ([published Online First: Epub Date]|).

120. Chu CT, Ji J, Dagda RK, et al. Cardiolipin externalization to the outer mitochondrial membrane acts as an elimination signal for mitophagy in neuronal cells. *Nat Cell Biol.* 2013;15(10):1197−1205. https://doi.org/10.1038/ncb2837 ([published Online First: Epub Date]|).

121. Wan M, Hua X, Su J, et al. Oxidized but not native cardiolipin has pro-inflammatory effects, which are inhibited by Annexin A5. *Atherosclerosis.* 2014;235(2):592−598. https://doi.org/10.1016/j.atherosclerosis.2014.05.913 ([published Online First: Epub Date]|).

122. Wiley CD, Velarde MC, Lecot P, et al. Mitochondrial dysfunction induces senescence with a distinct secretory phenotype. *Cell Metabol.* 2016;23(2):303−314. https://doi.org/10.1016/j.cmet.2015.11.011 ([published Online First: Epub Date]|).

123. Freund A, Laberge RM, Demaria M, Campisi J. Lamin B1 loss is a senescence-associated biomarker. *Mol Biol Cell.* 2012;23(11):2066−2075. https://doi.org/10.1091/mbc.E11-10-0884 ([published Online First: Epub Date]|).

124. Davalos AR, Kawahara M, Malhotra GK, et al. p53-dependent release of Alarmin HMGB1 is a central mediator of senescent phenotypes. *J Cell Biol.* 2013;201(4):613−629. https://doi.org/10.1083/jcb.201206006 ([published Online First: Epub Date]|).

125. Houtkooper RH, Cantó C, Wanders RJ, Auwerx J. The secret life of NAD+: an old metabolite controlling new metabolic signaling pathways. *Endocr Rev.* 2010;31(2):194−223. https://doi.org/10.1210/er.2009-0026 ([published Online First: Epub Date]|).

126. Lee SM, Dho SH, Ju SK, Maeng JS, Kim JY, Kwon KS. Cytosolic malate dehydrogenase regulates senescence in human fibroblasts. *Biogerontology.* 2012;13(5):525−536. https://doi.org/10.1007/s10522-012-9397-0 ([published Online First: Epub Date]|).

127. Braidy N, Guillemin GJ, Mansour H, Chan-Ling T, Poljak A, Grant R. Age related changes in NAD+ metabolism oxidative stress and Sirt1 activity in wistar rats. *PloS One.* 2011;6(4):e19194. https://doi.org/10.1371/journal.pone.0019194 ([published Online First: Epub Date]|).

128. Gomes AP, Price NL, Ling AJ, et al. Declining NAD(+) induces a pseudohypoxic state disrupting nuclear-mitochondrial communication during aging. *Cell.* 2013;155(7):1624−1638. https://doi.org/10.1016/j.cell.2013.11.037 ([published Online First: Epub Date]|).

129. Stein LR, Imai S. Specific ablation of Nampt in adult neural stem cells recapitulates their functional defects during aging. *Embo J.* 2014;33(12):1321−1340. https://doi.org/10.1002/embj.201386917 ([published Online First: Epub Date]|).

130. Yoshino J, Mills KF, Yoon MJ, Imai S. Nicotinamide mononucleotide, a key NAD(+) intermediate, treats the pathophysiology of diet- and age-induced diabetes in mice. *Cell Metabol.* 2011;14(4):528−536. https://doi.org/10.1016/j.cmet.2011.08.014 ([published Online First: Epub Date]|).

131. Aman Y, Schmauck-Medina T, Hansen M, et al. Autophagy in healthy aging and disease. *Nat Aging*. 2021;1(8):634–650. https://doi.org/10.1038/s43587-021-00098-4 ([published Online First: Epub Date]|).

132. Tai H, Wang Z, Gong H, et al. Autophagy impairment with lysosomal and mitochondrial dysfunction is an important characteristic of oxidative stress-induced senescence. *Autophagy*. 2017;13(1):99–113. https://doi.org/10.1080/15548627.2016.1247143 ([published Online First: Epub Date]|).

133. Kang C, Xu Q, Martin TD, et al. The DNA damage response induces inflammation and senescence by inhibiting autophagy of GATA4. *Science*. 2015;349(6255):aaa5612. https://doi.org/10.1126/science.aaa5612 ([published Online First: Epub Date]|).

134. Kang HT, Lee KB, Kim SY, Choi HR, Park SC. Autophagy impairment induces premature senescence in primary human fibroblasts. *PloS One*. 2011;6(8):e23367. https://doi.org/10.1371/journal.pone.0023367 ([published Online First: Epub Date]|).

135. Correia-Melo C, Birch J, Fielder E, et al. Rapamycin improves healthspan but not inflammaging in nfkb1(-/-) mice. *Aging Cell*. 2019;18(1):e12882. https://doi.org/10.1111/acel.12882 ([published Online First: Epub Date]|).

136. Correia-Melo C, Marques FD, Anderson R, et al. Mitochondria are required for pro-ageing features of the senescent phenotype. *Embo J*. 2016;35(7):724–742. https://doi.org/10.15252/embj.201592862 ([published Online First: Epub Date]|).

137. Lou G, Palikaras K, Lautrup S, Scheibye-Knudsen M, Tavernarakis N, Fang EF. Mitophagy and neuroprotection. *Trends Mol Med*. 2020;26(1):8–20. https://doi.org/10.1016/j.molmed.2019.07.002 ([published Online First: Epub Date]|).

138. Fivenson EM, Lautrup S, Sun N, et al. Mitophagy in neurodegeneration and aging. *Neurochem Int*. 2017;109:202–209. https://doi.org/10.1016/j.neuint.2017.02.007 ([published Online First: Epub Date]|).

139. Ahmad T, Sundar IK, Lerner CA, et al. Impaired mitophagy leads to cigarette smoke stress-induced cellular senescence: implications for chronic obstructive pulmonary disease. *Faseb J*. 2015;29(7):2912–2929. https://doi.org/10.1096/fj.14-268276 ([published Online First: Epub Date]|).

140. Hoshino A, Mita Y, Okawa Y, et al. Cytosolic p53 inhibits Parkin-mediated mitophagy and promotes mitochondrial dysfunction in the mouse heart. *Nat Commun*. 2013;4:2308. https://doi.org/10.1038/ncomms3308 ([published Online First: Epub Date]|).

141. Manzella N, Santin Y, Maggiorani D, et al. Monoamine oxidase-A is a novel driver of stress-induced premature senescence through inhibition of parkin-mediated mitophagy. *Aging Cell*. 2018;17(5):e12811. https://doi.org/10.1111/acel.12811 ([published Online First: Epub Date]|).

142. Serrano M, Lin AW, McCurrach ME, Beach D, Lowe SW. Oncogenic ras provokes premature cell senescence associated with accumulation of p53 and p16INK4a. *Cell*. 1997;88(5):593–602. https://doi.org/10.1016/s0092-8674(00)81902-9 ([published Online First: Epub Date]|).

143. Muñoz-Espín D, Serrano M. Cellular senescence: from physiology to pathology. *Nat Rev Mol Cell Biol*. 2014;15(7):482–496. https://doi.org/10.1038/nrm3823 ([published Online First: Epub Date]|).

144. Liu XL, Ding J, Meng LH. Oncogene-induced senescence: a double edged sword in cancer. *Acta Pharmacol Sin*. 2018;39(10):1553–1558. https://doi.org/10.1038/aps.2017.198 ([published Online First: Epub Date]|).

145. Kruse JP, Gu W. Modes of p53 regulation. *Cell*. 2009;137(4):609–622. https://doi.org/10.1016/j.cell.2009.04.050 ([published Online First: Epub Date]|).

146. Bartkova J, Rezaei N, Liontos M, et al. Oncogene-induced senescence is part of the tumorigenesis barrier imposed by DNA damage checkpoints. *Nature*.

2006;444(7119):633–637. https://doi.org/10.1038/nature05268 ([published Online First: Epub Date]|).

147. Di Micco R, Fumagalli M, Cicalese A, et al. Oncogene-induced senescence is a DNA damage response triggered by DNA hyper-replication. *Nature.* 2006;444(7119):638–642. https://doi.org/10.1038/nature05327 ([published Online First: Epub Date]|).

148. Bartkova J, Horejsí Z, Koed K, et al. DNA damage response as a candidate anti-cancer barrier in early human tumorigenesis. *Nature.* 2005;434(7035):864–870. https://doi.org/10.1038/nature03482 ([published Online First: Epub Date]|).

149. Trost TM, Lausch EU, Fees SA, et al. Premature senescence is a primary fail-safe mechanism of ERBB2-driven tumorigenesis in breast carcinoma cells. *Cancer Res.* 2005;65(3):840–849.

150. Moiseeva O, Mallette FA, Mukhopadhyay UK, Moores A, Ferbeyre G. DNA damage signaling and p53-dependent senescence after prolonged beta-interferon stimulation. *Mol Biol Cell.* 2006;17(4):1583–1592. https://doi.org/10.1091/mbc.e05-09-0858 ([published Online First: Epub Date]|).

151. Alimonti A, Nardella C, Chen Z, et al. A novel type of cellular senescence that can be enhanced in mouse models and human tumor xenografts to suppress prostate tumorigenesis. *J Clin Invest.* 2010;120(3):681–693. https://doi.org/10.1172/jci40535 ([published Online First: Epub Date]|).

152. Prieur A, Peeper DS. Cellular senescence in vivo: a barrier to tumorigenesis. *Curr Opin Cell Biol.* 2008;20(2):150–155. https://doi.org/10.1016/j.ceb.2008.01.007 ([published Online First: Epub Date]|).

153. Osaki M, Oshimura M, Ito H. PI3K-Akt pathway: its functions and alterations in human cancer. *Apoptosis : An Int J Program Cell Death.* 2004;9(6):667–676. https://doi.org/10.1023/B:APPT.0000045801.15585 (dd[published Online First: Epub Date]|).

154. Alimonti A. PTEN breast cancer susceptibility: a matter of dose. *Ecancermedicalscience.* 2010;4:192. https://doi.org/10.3332/ecancer.2010.192 ([published Online First: Epub Date]|).

155. Chen Z, Trotman LC, Shaffer D, et al. Crucial role of p53-dependent cellular senescence in suppression of Pten-deficient tumorigenesis. *Nature.* 2005;436(7051):725–730. https://doi.org/10.1038/nature03918 ([published Online First: Epub Date]|).

156. Majumder PK, Febbo PG, Bikoff R, et al. mTOR inhibition reverses Akt-dependent prostate intraepithelial neoplasia through regulation of apoptotic and HIF-1-dependent pathways. *Nat Med.* 2004;10(6):594–601. https://doi.org/10.1038/nm1052 ([published Online First: Epub Date]|).

157. Podsypanina K, Lee RT, Politis C, et al. An inhibitor of mTOR reduces neoplasia and normalizes p70/S6 kinase activity in Pten+/- mice. *Proc Nat Acad Sci U S A.* 2001;98(18):10320–10325. https://doi.org/10.1073/pnas.171060098 ([published Online First: Epub Date]|).

158. Cheung TH, Rando TA. Molecular regulation of stem cell quiescence. *Nat Rev Mol Cell Biol.* 2013;14(6):329–340. https://doi.org/10.1038/nrm3591 ([published Online First: Epub Date]|).

159. Mushtaq M, Kovalevska L, Darekar S, et al. Cell stemness is maintained upon concurrent expression of RB and the mitochondrial ribosomal protein S18-2. *Proc Nat Acad Sci U S A.* 2020;117(27):15673–15683. https://doi.org/10.1073/pnas.192253 5117 ([published Online First: Epub Date]|).

160. Tower J. Stress and stem cells. *Wiley Interdiscip Rev Dev Biol.* 2012;1(6):789–802. https://doi.org/10.1002/wdev.56 ([published Online First: Epub Date]|).

161. Oh J, Lee YD, Wagers AJ. Stem cell aging: mechanisms, regulators and therapeutic opportunities. *Nat Med.* 2014;20(8):870–880. https://doi.org/10.1038/nm.3651 ([published Online First: Epub Date]|).

162. Warren LA, Rossi DJ. Stem cells and aging in the hematopoietic system. *Mech Ageing Dev.* 2009;130(1–2):46–53. https://doi.org/10.1016/j.mad.2008.03.010 ([published Online First: Epub Date]|).

163. Sousa-Victor P, Gutarra S, García-Prat L, et al. Geriatric muscle stem cells switch reversible quiescence into senescence. *Nature.* 2014;506(7488):316–321. https://doi.org/10.1038/nature13013 ([published Online First: Epub Date]|).

164. Boyle M, Wong C, Rocha M, Jones DL. Decline in self-renewal factors contributes to aging of the stem cell niche in the Drosophila testis. *Cell Stem Cell.* 2007;1(4):470–478. https://doi.org/10.1016/j.stem.2007.08.002 ([published Online First: Epub Date]|).

165. Rossi DJ, Bryder D, Seita J, Nussenzweig A, Hoeijmakers J, Weissman IL. Deficiencies in DNA damage repair limit the function of haematopoietic stem cells with age. *Nature.* 2007;447(7145):725–729. https://doi.org/10.1038/nature05862 ([published Online First: Epub Date]|).

166. Folmes CD, Nelson TJ, Martinez-Fernandez A, et al. Somatic oxidative bioenergetics transitions into pluripotency-dependent glycolysis to facilitate nuclear reprogramming. *Cell Metabol.* 2011;14(2):264–271. https://doi.org/10.1016/j.cmet.2011.06.011 ([published Online First: Epub Date]|).

167. Chandel NS, Jasper H, Ho TT, Passegué E. Metabolic regulation of stem cell function in tissue homeostasis and organismal ageing. *Nat Cell Biol.* 2016;18(8):823–832. https://doi.org/10.1038/ncb3385 ([published Online First: Epub Date]|).

168. Takahashi K, Yamanaka S. A decade of transcription factor-mediated reprogramming to pluripotency. *Nat Rev Mol Cell Biol.* 2016;17(3):183–193. https://doi.org/10.1038/nrm.2016.8 ([published Online First: Epub Date]|).

169. Al-Nbaheen M, Vishnubalaji R, Ali D, et al. Human stromal (mesenchymal) stem cells from bone marrow, adipose tissue and skin exhibit differences in molecular phenotype and differentiation potential. *Stem Cell Rev Rep.* 2013;9(1):32–43. https://doi.org/10.1007/s12015-012-9365-8 ([published Online First: Epub Date]|).

170. Liu J, Ding Y, Liu Z, Liang X. Senescence in mesenchymal stem cells: functional alterations, molecular mechanisms, and rejuvenation strategies. *Front Cell Develop Biol.* 2020;8. https://doi.org/10.3389/fcell.2020.00258 ([published Online First: Epub Date]|).

171. Banimohamad-Shotorbani B, Kahroba H, Sadeghzadeh H, et al. DNA damage repair response in mesenchymal stromal cells: from cellular senescence and aging to apoptosis and differentiation ability. *Ageing Res Rev.* 2020;62:101125. https://doi.org/10.1016/j.arr.2020.101125 ([published Online First: Epub Date]|).

172. Gnani D, Crippa S, Della Volpe L, et al. An early-senescence state in aged mesenchymal stromal cells contributes to hematopoietic stem and progenitor cell clonogenic impairment through the activation of a pro-inflammatory program. *Aging Cell.* 2019;18(3):e12933. https://doi.org/10.1111/acel.12933 ([published Online First: Epub Date]|).

173. Ellison-Hughes GM. First evidence that senolytics are effective at decreasing senescent cells in humans. *EBioMedicine.* 2020;56:102473. https://doi.org/10.1016/j.ebiom.2019.09.053 ([published Online First: Epub Date]|).

174. Kirkland JL, Tchkonia T, Zhu Y, Niedernhofer LJ, Robbins PD. The clinical potential of senolytic drugs. *J Am Geriatr Soc.* 2017;65(10):2297–2301. https://doi.org/10.1111/jgs.14969 ([published Online First: Epub Date]|).

175. Tchkonia T, Kirkland JL. Aging, cell senescence, and chronic disease: emerging therapeutic strategies. *Jama.* 2018;320(13):1319–1320. https://doi.org/10.1001/jama.2018.12440 ([published Online First: Epub Date]|).

176. Yousefzadeh MJ, Zhu Y, McGowan SJ, et al. Fisetin is a senotherapeutic that extends health and lifespan. *EBioMedicine.* 2018;36:18–28. https://doi.org/10.1016/j.ebiom.2018.09.015 ([published Online First: Epub Date]|).

177. Justice JN, Nambiar AM, Tchkonia T, et al. Senolytics in idiopathic pulmonary fibrosis: results from a first-in-human, open-label, pilot study. *EBioMedicine.* 2019;40:554–563. https://doi.org/10.1016/j.ebiom.2018.12.052 ([published Online First: Epub Date]|).

178. Niedernhofer LJ, Robbins PD. Senotherapeutics for healthy ageing. *Nat Rev Drug Dis.* 2018;17(5):377. https://doi.org/10.1038/nrd.2018.44 ([published Online First: Epub Date]|).

179. Tilstra JS, Robinson AR, Wang J, et al. NF-κB inhibition delays DNA damage–induced senescence and aging in mice. *J Clin Invest.* 2012;122(7):2601–2612. https://doi.org/10.1172/JCI45785 ([published Online First: Epub Date]|).

180. Kohli J, Wang B, Brandenburg SM, et al. Algorithmic assessment of cellular senescence in experimental and clinical specimens. *Nat Protoc.* 2021;16(5):2471–2498. https://doi.org/10.1038/s41596-021-00505-5 ([published Online First: Epub Date]|).

181. Cho KA, Ryu SJ, Oh YS, et al. Morphological adjustment of senescent cells by modulating caveolin-1 status. *J Biol Chem.* 2004;279(40):42270–42278. https://doi.org/10.1074/jbc.M402352200 ([published Online First: Epub Date]|).

182. Michishita E, Nakabayashi K, Suzuki T, et al. 5-Bromodeoxyuridine induces senescence-like phenomena in mammalian cells regardless of cell type or species. *J Biochem.* 1999;126(6):1052–1059. https://doi.org/10.1093/oxfordjournals.jbchem.a022549 ([published Online First: Epub Date]|).

183. Powell CD, Van Zandycke SM, Quain DE, Smart KA. Replicative ageing and senescence in *Saccharomyces cerevisiae* and the impact on brewing fermentations. *Microbiology (Reading).* 2000;146(Pt 5):1023–1034. https://doi.org/10.1099/00221287-146-5-1023 ([published Online First: Epub Date]|).

184. Gonzalez-Gualda E, Baker AG, Fruk L, Munoz-Espin D. A guide to assessing cellular senescence in vitro and in vivo. *FEBS J.* 2021;288(1):56–80. https://doi.org/10.1111/febs.15570 ([published Online First: Epub Date]|).

185. Sherr CJ. Ink4-Arf locus in cancer and aging. *Wiley Interdiscip Rev Dev Biol.* 2012;1(5):731–741. https://doi.org/10.1002/wdev.40 ([published Online First: Epub Date]|).

186. Blagosklonny MV. Cell cycle arrest is not senescence. *Aging.* 2011;3(2):94–101. https://doi.org/10.18632/aging.100281 ([published Online First: Epub Date]|).

187. Cuollo L, Antonangeli F, Santoni A, Soriani A. The senescence-associated secretory phenotype (SASP) in the challenging future of cancer therapy and age-related diseases. *Biology.* 2020;9(12). https://doi.org/10.3390/biology9120485 ([published Online First: Epub Date]|).

188. Di Micco R, Sulli G, Dobreva M, et al. Interplay between oncogene-induced DNA damage response and heterochromatin in senescence and cancer. *Nat Cell Biol.* 2011;13(3):292–302. https://doi.org/10.1038/ncb2170 ([published Online First: Epub Date]|).

189. Zhang R, Chen W, Adams PD. Molecular dissection of formation of senescence-associated heterochromatin foci. *Mol Cell Biol.* 2007;27(6):2343–2358. https://doi.org/10.1128/MCB.02019-06 ([published Online First: Epub Date]|).

190. Montpetit AJ, Alhareeri AA, Montpetit M, et al. Telomere length: a review of methods for measurement. *Nurs Res.* 2014;63(4):289–299. https://doi.org/10.1097/NNR.0000000000000037 ([published Online First: Epub Date]|).

191. Brunk UT, Terman A. Lipofuscin: mechanisms of age-related accumulation and influence on cell function. *Free Radic Biol Med*. 2002;33(5):611—619. https://doi.org/10.1016/s0891-5849(02)00959-0 ([published Online First: Epub Date]|).
192. http://www.saspatlas.com/.
193. Basisty N, Kale A, Jeon OH, et al. A proteomic atlas of senescence-associated secretomes for aging biomarker development. *PLoS Biol*. 2020;18(1):e3000599. https://doi.org/10.1371/journal.pbio.3000599 ([published Online First: Epub Date]|).
194. Noren Hooten N, Evans MK. Techniques to induce and quantify cellular senescence. *J Vis Exp*. 2017;123. https://doi.org/10.3791/55533 ([published Online First: Epub Date]|).
195. Vasileiou PVS, Evangelou K, Vlasis K, et al. Mitochondrial homeostasis and cellular senescence. *Cells*. 2019;8(7). https://doi.org/10.3390/cells8070686 ([published Online First: Epub Date]|).
196. Ang J, Lee YA, Raghothaman D, et al. Rapid detection of senescent mesenchymal stromal cells by a fluorescent probe. *Biotechnol J*. 2019;14(10):e1800691. https://doi.org/10.1002/biot.201800691 ([published Online First: Epub Date]|).
197. Victorelli S, Passos JF. Reactive oxygen species detection in senescent cells. *Meth Mol Biol*. 2019;1896:21—29. https://doi.org/10.1007/978-1-4939-8931-7_3 ([published Online First: Epub Date]|).
198. Canto C, Menzies KJ, Auwerx J. NAD(+) metabolism and the control of energy homeostasis: a balancing act between mitochondria and the nucleus. *Cell Metabol*. 2015;22(1):31—53. https://doi.org/10.1016/j.cmet.2015.05.023 ([published Online First: Epub Date]|).
199. Ogrodnik M. Cellular aging beyond cellular senescence: markers of senescence prior to cell cycle arrest in vitro and in vivo. *Aging Cell*. 2021;20(4):e13338. https://doi.org/10.1111/acel.13338 ([published Online First: Epub Date]|).
200. Childs BG, Durik M, Baker DJ, van Deursen JM. Cellular senescence in aging and age-related disease: from mechanisms to therapy. *Nat Med*. 2015;21(12):1424—1435. https://doi.org/10.1038/nm.4000 (published Online First: Epub Date]|).
201. Gorgoulis V, Adams PD, Alimonti A, et al. Cellular senescence: defining a path forward. *Cell*. 2019;179(4):813—827. https://doi.org/10.1016/j.cell.2019.10.005 ([published Online First: Epub Date]|).

SECTION V

Applications

CHAPTER 7

Aging and the immune system

Wenliang Pan
Division of Rheumatology and Clinical Immunology, Department of Medicine, Beth Israel Deaconess Medical Center, Harvard Medical School, Boston, MA, United States

1. Overview of immunosenescence

The immune system is a complex network of cells and tissues working to maintain the health of an organism. It protects an organism against foreign pathogens and helps to maintain the homeostasis of the organism by eliminating dysfunctional or dead cells. Like any other system, the immune system changes with age, which is associated with an array of defects in both innate and adaptive immunity, collectively termed immunosenescence. Immunosenescence is a multifaceted phenomenon at the root of age-associated immune dysfunction. There is growing evidence that the hallmarks of aging, particularly the senescence of cells, are attributed to compromised organismal barriers, excessive inflammatory signaling, and exhaustion of both innate and adaptive immunity. These in turn result in a myriad of pathologic conditions including, but not limited to, the decreased ability to surveil and clear senescent and cancerous cells, an increased autoimmune response leading to tissue damage, the reduced ability to tackle pathogens, the decreased competence to elicit a robust response to vaccination, and contributions to high susceptibility to infectious, cardiovascular, and neurologic diseases, diabetes, and cancer. Owing to the critical role of immunity in sustaining organismal integrity and clearing away damaged cells, it is unsurprising that immunosenescence constitutes both a consequence of aging and a further accelerator of organismal functional decay.

2. Aging-related lymphoid organs and immunity

Immunosenescence, which is a process of immune dysfunction that occurs with age, stems from two complementary processes: (1) as the indirect consequence of tissue (lymphoid organ) cellular senescence, which deteriorates organismal barriers and promotes the release of inflammatory signaling molecules to which immune cells respond; and (2) as the direct effect of senescence of the immune cells, which contribute to innate and adaptive immunity.

Molecular, Cellular, and Metabolic Fundamentals of Human Aging
ISBN 978-0-323-91617-2
https://doi.org/10.1016/B978-0-323-91617-2.00003-1

© 2023 Elsevier Inc.
All rights reserved.

2.1 Lymphoid organs

2.1.1 Bone marrow

Bone marrow is a spongy tissue residing in the core of the vertebrae, skull, and long bones. The bone marrow microenvironment provides regions that support the function of hematopoietic stem cells (HSCs), which give rise to immune cells including myeloid and lymphoid lineages. Cells within these regions constitute the HSC niche. The HSC niche contributes to controlling HSC quiescence, proliferation, self-renewal, and differentiation. Accumulating evidence suggests that the aging of HSCs initiates aging of the immune system and directly contributes to immunosenescence. A set of criteria used to define the phenotype of aged HSCs includes[1] an increased number of HSCs; the reduced regenerative capacity of HSCs; skewed differentiation potential of HSCs; enhanced mobilization of HSCs from the bone marrow into the blood; and reduced homing of HSCs back to the bone marrow, with those that return home showing a distinct niche selectivity (distinct positions relative to the endosteal niche).

Increased hematopoietic stem cell number and decreased regenerative potential. The number of HSCs in the aged bone marrow increases by two- to tenfold in both mice and humans.[2–4] Instinctively, the increased HSC numbers observed in elderly individuals are considered to be beneficial, because these HSCs might, for instance, maintain immune functions. Nevertheless, under conditions of stress and regeneration, aged HSCs display functional defects such as diminished regenerative potential, because of their reduced long-term self-renewal capacity. The increase in the number of HSCs does not compensate for their loss in function, leading to an overall reduction in the regenerative capacity of the pool of aged HSCs.[5]

The underlying mechanisms by which the aging-associated increase in HSC numbers occur are unclear. The high frequency of self-renewing symmetric cell divisions might give the increased numbers and impaired function of aged HSCs.[3] In addition, a loss of responsiveness to extrinsic signals or age-related changes in the expression and secretion of systemic and niche factors has been proposed to cause the high numbers and functional defects of aged HSCs.[6] Alternatively, a cell–intrinsic mechanism such as mitochondrial aging is theorized to drive the changes in HSC numbers during aging. Moreover, increased expression of cyclin–dependent kinase inhibitor 2A and loss of acetylation of lysine 16 on the tail of histone H4 are seen in aged HSCs and are associated with their increased number and compromised self-renewal capacity.[3]

Skewed differentiation potential of hematopoietic stem cells. Lineage specification occurs during the early stages of hematopoiesis after the differentiation of HSCs into common lymphoid progenitor (CLP) cells and common myeloid progenitor (CMP) cells or after the separation of cells with megakaryocyte-erythroid potential from those with granulocyte-macrophage, B-cell, and T-cell potential. Aged HSCs exhibit a markedly decreased output of cells of the lymphoid and erythroid lineages, whereas the myeloid lineage output of aged HSCs is maintained or even increased compared with young HSCs.[3] Despite their increased numbers, the quality of myeloid cells produced by aged HSCs is compromized,[7] similar to what is observed with HSCs themselves. Early lymphoid to myeloid lineage skewing might be caused by changes in HSC differentiation or by the altered proliferation or survival of CMP cells and CLP cells. There is considerable functional heterogeneity among the most primitive HSCs in terms of their differentiation potentials. Some HSCs show a relatively low capacity to differentiate into lymphoid cells and can be regarded as myeloid-biased, some HSCs show the opposite behavior and can be regarded as lymphoid-biased, whereas other HSCs seem to maintain a balanced output and are regarded as balanced HSCs.[8] Consequently, an alternative explanation for aging-induced lineage skewing is that the differentiation potential of individual HSCs does not change with age; instead, the composition of the HSC pool is altered. At the single-cell level, epigenetic program deterioration with age has been considered a reason for HSCs giving rise to myeloid-biased stem cells upon aging.[9] In addition, it has been suggested that HSC niche-secreted chemokine ligand 5 increases with age in the bone marrow and can drive myeloid skewing of HSCs through decreasing prolymphoid transcription factors.[10] Age-related accumulation of adipocytes in the bone marrow has been attributed to the increased expression of receptor activator of nuclear factor κ-B ligand. These bone marrow adipocytes in turn produce an array of factors that have been shown to affect hematopoiesis and skew it toward myeloid lineage.[11]

Altered homing and mobilization of hematopoietic stem cells. HSC niches comprise multiple nonhematopoietic cells, such as endothelial cells, osteoblasts, adipocytes, and mesenchymal stem cells (MSCs), which promote HSC maintenance by regulating HSC self-renewal, quiescence, mobilization, and differentiation. Young HSCs are home to the bone marrow and localize in close proximity to the endosteal niche instead of the perivascular niche. However, the localization of aged HSCs is away from the endosteal stem cell niche in the bone marrow after their transplantation.[12] This implies that aged HSCs select niches that are distinct from those that young HSCs occupy. Aged HSCs show about a two-fold reduced ability to home to the bone marrow and higher mobilization in response to cytokine stimulation compared with young HSCs.[12] CXC-chemokine ligand 12 (CXCL12) is an essential chemokine produced by osteoblasts in the bone marrow niche and functions as a chemoattractant for HSCs to regulate their localization, turnover, and mobilization in the bone marrow. The altered mobilization of aged HSCs might be attributed to the decreased plasma levels of CXCL12 in the niche caused by increased adiposity and reduced osteogenesis upon aging.[11] In addition, the sympathetic nervous system, which regulates interactions between HSCs and MSCs, and shows altered activation status upon aging, might be a critical factor in regulating HSC mobilization.[13]

2.1.2 Thymus

The thymus, which is located behind the breastbone and above the heart in nearly all vertebrates, is a primary lymphoid organ essential for the maturation of bone marrow—derived T lymphoid precursors. In mammals, the thymus consists of two main anatomic sites, the cortex and the medulla, where T lymphocytes develop into a pool of circulating naive $CD4^+$, $CD8^+$, and regulatory T cells (Tregs) after undergoing positive, negative, and agonist selections. A proper thymic output is thus considered a crucial first stage that generates a broad but self-tolerant T-cell repertoire, which is vital to the development of a strong adaptive immune response against pathogens and tumors while minimizing the risk for autoimmune diseases.

Thymic involution upon aging. In an evolutionarily conserved manner, most vertebrates experience age-associated thymic involution beginning in childhood and peaking around puberty. Thymic involution is characterized by atrophy, disruption of tissue architecture, and a progressive decrease in thymic cellularity along with enlarged perivascular areas and accumulating adipocytes.[14,15] Interestingly, the recruitment of T-cell progenitors to the thymus is similar between young and old mice. In contrast, dynamic involution results in a decrease in the thymic output of undifferentiated T cells (e.g., naive T cells) but an increase in terminally differentiated cells (e.g., T cells with memory phenotypes), which causes a constriction in the diversity of peripheral T cell receptor (TCR) repertoire.[16] However, even naive T cells that are produced in the aged thymus are functionally impaired, expressing higher levels of senescence markers such as CD57 and exhibiting limited proliferation in response to antigen stimulation.[17] Furthermore, they have reduced homing receptors such as CD62L and CCR7, interfering with their mobilization to relevant sites of action.[18] The overall consequence of these changes is a defective adaptive immune response to neoantigens, including new pathogens, vaccinations, and tumor-associated antigens.

Factors that induce thymic involution. The thymic output is coordinated by both the stromal (e.g., cortex and medulla) and lymphoid compartments. Architecturally, the nonhematopoietic stromal microenvironment of the thymus undergoes marked structural changes with age, primarily within the thymus epithelial cell (TEC) compartment. With age, the cortical and medullary thymic epithelial regions become irregular and atrophied, with loss of the definition of the corticomedullary junction, diminution of the thymic epithelial space, enlargement of the perivascular space, and increased adiposity. Mechanistically, several factors regulate these compartments and affect the involution process. The sex hormones testosterone and estrogen, which are increased during aging, facilitate the involution process mainly by affecting the TECs, and, to a lesser extent, via direct signaling in thymocytes.[19] Inflammation and stress, which activate the hypothalamus—pituitary—adrenal axis and induce the production of cortisol, have a role in thymus functionality owing to the abundant expression of the glucocorticoid receptor on thymocytes.[20] In addition, the reduced production of keratinocyte growth factor (KGF), prolongevity factor fibroblast growth factor (FGF) 21, growth hormone (GH), insulin-like growth factor-1 (IGF-1), transcription factor Forkhead box (FOX) N1, and cytokines interleukin (IL)-7 and IL-22 contributes to age-related thymic involution.[15] Specifically, KGF, which is primarily secreted in a paracrine fashion from mesenchymal cells, binds to an epithelial cell—specific splice variant of fibroblast growth factor receptor-2 expressed on TECs within the thymus and promotes TEC proliferation and thymopoiesis. FGF21 and its receptor are expressed in thymic stromal cells and have a key role in maintaining the levels of thymic progenitors, cortical epithelial cells, and circulating naive T cells in aging. GH and IGF-1 are expressed by the thymic stromal compartment, in particular by TECs and fibroblasts, whereas their receptors are expressed not only by the stromal compartment (mainly by the epithelial cells and fibroblasts) but also by the hematopoietic compartment. Administration of GH improves thymic cellularity, increases TCR diversity, and enhances recovery of the hematopoietic compartment. The FOXN1 transcription factor is expressed in TECs and regulates their development and function (e.g., antigen presentation and T-cell selection). Its downregulation with aging acceler-ates thymic atrophy, a process that is at least partly regulated by bone morphogenetic protein 4 expressed by TECs. The IL-7 cytokine, produced mainly by TECs, is involved in the migration, maturation, and maintenance of lymphocytes within the thymus. Reduced levels of IL-22 with age are implicated in dysfunctional TECs and reduced thymopoiesis. The accu-mulation of abnormal mesenchymal cells may also lead to abnormal signaling, such as adipokine secretion by adipocytes, which may further interfere with thymopoiesis.

2.1.3 Spleen

The spleen is a secondary lymphoid organ organized into regions called the red pulp and white pulp, which are separated by an interface called the marginal zone (MZ). Blood circulation in the spleen is open: afferent arterial blood ends in sinusoids at the MZ that surround the white pulp. Blood flows through sinusoid spaces and red pulp into venous sinuses, which collect into efferent splenic veins. The splenic red pulp serves mostly to filter blood and recycle iron from aging red blood cells. Diverse splenic populations trap and remove blood-borne antigens as well as initiate innate and adaptive immune responses against pathogens. For instance, the white pulp hosts T-cell and B-cell zones and allows the generation of antigen-specific immune responses that protect the body against foreign patho-gens. In addition, the red pulp and MZ contain all major types of mono-nuclear phagocytes, including macrophages, dendritic cells (DCs), and monocytes. These cells constitute the line of protection of the organism because they identify pathogens and cellular stress, eradicate dying cells and foreign materials, maintain tissue homeostasis and inflammatory responses, and shape adaptive immunity.[21]

Altered marginal zone structure and function with age. Upon aging, The MZ region of the murine spleen undergoes significant structural disorganization. MZ macrophages have altered distribution, no longer forming a continuous boundary along the MZ. The MAdCAM-1$^+$ MZ sinus-lining cells in aged murine spleens also no longer form a continuous boundary between the follicle and MZ and become thicker in density. The distribution and density of the marginal metallophilic macrophages in the inner layer of the MZ are also disturbed in aged murine spleens. Func-tionally, the phagocytic capacity of MZ macrophages from aged BALB/c mice exhibits less efficiency in vivo, although no difference in phagocytosis between young and aged MZ macrophages was observed under in vitro conditions.[22]

Altered white pulp structure and function with age. Upon aging, the microarchitecture and cellularity of the spleen significantly change, accompanying by altered localization of various cells. In the aged murine spleen, the T- and B-cell regions within the white pulp decompartmentalize and the boundaries become obscure.[23] Immune responses such as immunization with T-independent and T-dependent antigens within spleens are often associated with phases of tissue remodeling that are later resolved in young mice. However, aged spleens are unable to resolve this structural disorganization. Furthermore, the aged splenic microenvironment adversely affects the migration of immature and MZ B cells within it. Stromal cells in the spleen have key roles in coordinating the localization of leukocytes to specific tissue niches. The disorganized compartmentalization of the T- and B-cell regions is most likely due to aging effects on the underlying stromal cells within these zones, which also become significantly altered in their distribution.[22]

2.1.4 Lymph nodes

Lymph nodes (LNs) are small bulbous structures that form a crucial part of the lymphatic system along with the lymphatic vessels (LVs). They filter the lymph fluid obtained from surrounding tissues before it reenters the bloodstream. LNs house various immune cells including T cells, B cells, and DCs and have an essential role in establishing a strong immune response. The primary function of the LN is to coordinate immune responses to antigens trafficking from peripheral tissues. The nonhematopoietic stromal cell subsets provide the architecture and scaffolding necessary to guide cellular trafficking and compartmentalization, facilitate antigen presentation to circulating naive T and B cells, and thus promote immune surveillance against infection.[24] In addition, LN stromal cells are responsible for the production and presentation of chemokines that coordinate this trafficking of lymphocytes into and throughout the LNs.[24]

Chronic and progressive changes occur in LN with age. In general, LNs in both mice and humans become smaller and less cellular upon aging. The structural and functional disorganization of LNs that occurs with aging may adversely affect these processes.[25,26] Histologic evaluation of the structure of mesenteric LNs in the elderly has revealed general fibrosis, accumulation of adipocytes, thickening of the capsule and trabeculae, and increased amounts of connective tissue around the blood vessels.[25,26] The contribution of LN stromal cells to both immune homeostasis and function is evident. The dysfunction and/or disorganization of LN stromal cells, which is considered to contribute to immune senescence, have been observed during aging. The aged stromal cells in LNs associate with subpopulations including LVs, lymphatic endothelial cells (LECs), blood endothelial cells (BECs), high endothelial venule cells (HEVs), fibroblastic reticular cells (FRCs), and follicular DCs (FDCs).[25,26]

Lymphatic vessels and lymphatic endothelial cells in aged lymph nodes. Afferent LVs function as conduits for the trafficking of tissue fluid, macromolecules, antigens, bacteria, viruses, and immune cells from tissue to LNs. LECs line the sinuses of the LNs, supply antigens from the tissues, and allow cells to move to other nodes. As a semipermeable barrier, LECs sort the lymph-borne antigens into the LN parenchyma and lymphocytes exiting the parenchyma migrate through the LEC layer. Although no differences in the number of LECs in the LNs of old rodents were observed compared with young ones, aged mice show a diminished capacity to transport bacteria from peripheral tissues into the draining LN, probably owing to both increased LV permeability and reduced contractility of the musculature that surrounds the LVs.[27] Ultrastructural, biochemical, and functional comparative analyses indicate that elderly mice show a loss of extracellular matrix (ECM) proteins with fewer and more scattered collagen bundles typically surrounding endothelial cells and a reduced number of smooth muscle cells in the LVs.[27] Accordingly, the decreased ability to transport bacteria to the draining LNs and their tissue retention contribute to the reduced ability of the immune system to clear pathogens in the elderly.[27]

Blood endothelial cells and high endothelial venule cells in aged lymph nodes. HEVs are a critical subpopulation of BECs lining the postcapillary. HEVs that enable recirculation of naive T and B lymphocytes and central memory cells in various lymphoid organs ensure effective immune surveillance. HEVs have a cuboidal shape and a polarized expression of adhesion molecules so that circulating cells in the blood can anchor to the HEVs and extravasate into the LN. BECs (including HEVs) appear to be unchanged numerically in aged mice. However, aged HEVs appear to have a more dense and compressed morphology and exhibit increased permeability, inflammation, and number of senescent endothelial cells.[28] Moreover, aged HEVs may poorly facilitate lymphocyte entry into the aged LN. Therefore, the decrease in the number of postcapillary vessels and changes in the morphology and functionality of HEV cells that occur with age reduce the migration of cells to aging nodes, contributing to their cellular disorganization. All of these deficits make it difficult to initiate an early adaptive immune response, and thus contribute to the susceptibility of the aged organism to infections.[29]

Fibroblastic reticular cells in aged lymph nodes. FRCs constitute the main subpopulation of LN stromal cells and form a rigid three-dimensional network useful for the migration, accumulation, and accommodation of lymphocytes inside the parenchyma of the LN. FRCs have a significant role in organizing the T-cell zone within the LN and maintaining naive T-cell viability and function by producing IL-7 and the CCL19 chemokine. Altered IL-7 expression by FRCs in aged mouse LNs has been demonstrated and paralleled the decreased proliferation of the naive lymphocyte population.[28] Moreover, FRCs of old mice produce fewer CCL19 and CCL21 homeostatic chemokines and are less sensitive to antigenic challenges than young adult mice cells.[30] In addition, a reduced number of FRCs and their simultaneous densification cause disorganization of the FRC network in old LNs. The clear boundary between the cortical (B cells) and paracortical (T cells) zones in the mesenteric LNs has been blurred in aged mice.[31] As a consequence, the architecture of the lymphoid follicles is disturbed. The stroma of the cortical and paracortical zones appears more compressed and less reticulated. In addition, FRCs are an important source of LN collagen, an essential component of collagen fibers that determine the correct architecture and function of LNs. However, as a result of chronic inflammation caused by cytokines such as transforming growth factor (TGF)-β and IL-13, fibrosis occurs in aged regional LNs.[32] Moreover, FRCs are highly reactive and plastic, which facilitates their change within hours after vaccination and their increase in number during the swelling of LNs typical of the first few days of an immune response. However, the number of FRCs in old LNs increases slightly and does not reach the level observed in young adult LNs during infection. In addition, old FRCs are less stretchy. These phenomena limit the LN's ability to expand and accommodate the lymph draining the site of infection. Thus, they also limit antigen presentation to lymphocytes and reduce the influx of immune cells.[32,33]

Follicular dendritic cells in aged lymph nodes. FDCs are specialized cells of mesenchymal origin, named for their long cytoplasmic dendritic processes, and are unrelated to classical hematopoietic DC. During infection, FDCs support B-cell movement and proper localization to germinal centers (GCs) by producing CXCL13, B-cell activating factor of the tumor necrosis factor (TNF) family, and a proliferation-inducing ligand. Defects in FDCs are a potential factor contributing to poor humoral immunity in the elderly.[28] Aged FDCs contribute to lower affinity and impaired function antibody responses, fewer B cells within the LNs, less defined B-cell localization, and the reduced formation and size of GCs. The decreased area of FDCs and less CXCL13 protein produced by FDCs are found in LNs from aged mice compared with their young counterparts.[28] A reduction in the number and size of FDCs in old LNs compared with young LNs might impair the acquisition and retention of immunocomplexes in the LNs of old mice.[28] The aging of marginal reticular cells (MRCs) is responsible for defects in the MRC-to-FDC differentiation pathway.[28] Collectively, age-associated impairments in FDCs in LNs are a contributing factor to poor antigen retention, impaired GC formation, and a decline in the production of high-affinity, functionally neutralizing antibodies to infection and immunization.[28]

2.2 Immunity

The immune system may be schematically classified into two parts: (1) an ancestral/innate part constituting neutrophils, natural killer (NK), monocytes/macrophages, and DCs; and (2) a phylogenetically recent/adaptive part consisting of B and T lymphocytes.

2.2.1 Innate immunity

Upon aging, the innate immune system becomes dysregulated and innate immune cell functions required to respond to pathogens or vaccines are impaired in aged individuals. Dysfunction of the innate immune system is characterized by persistent inflammatory responses that involve multiple immune and nonimmune cell types, which vary depending on the cell activation state and tissue context.

2.2.1.1 Neutrophils

Neutrophils are the most abundant leukocytes in the blood. They represent the first line of defense of the innate immune response and kill invading microbes. They are the first cells to reach sites of infection to carry out their microbicidal activity rapidly, which relies on several mechanisms such as phagocytosis, degranulation of antimicrobial proteins, and the release of neutrophil extracellular traps (NETs).

Although there are no significant differences in the numbers of circulating neutrophils of healthy older people compared with younger subjects, their function is compromised in the elderly population.[34,35] The microbicidal activity of neutrophils from elderly individuals is significantly reduced owing to impaired phagocytosis, degranulation, and reactive oxygen species (ROS) production. Elderly individuals also display a reduced capability for NET formation owing to the increased release of neutrophil elastase via degranulation, an enzyme critical for NET formation. Moreover, a diminished respiratory burst of neutrophils from elderly subjects caused by diminished NADPH oxidase and myeloperoxidase activity can provide an additional explanation.[34,35] Decreased NETosis is frequently associated with sepsis, which explains why elderly individuals are more susceptible to invasive bacterial disease after skin and soft tissue infection. In addition, a potential mechanism contributing to an age-associated defect in intracellular bacteria killing in human neutrophils could involve the diminished expression of dehydroepiandrosterone sulfate, a circulating steroid that promotes superoxide generation in neutrophils.[34,35]

Age-associated defects in neutrophil effector function have been associated with impaired signal transduction functions.[35] Examples include impaired antiapoptotic responses to granulocyte-macrophage colony-stimulating factor (GM-CSF) mediated through Janus kinase (JAK)-signal transducer and activator of transcription (STAT) tyrosine kinase and the phosphoinositide 3-kinase-AKT pathways. Neutrophils from older adults compared with young individuals show reduced basal expression of suppressor of cytokine signaling 1 and 3, which negatively regulate JAK-STAT signaling, and an impaired response to triggering receptor expressed on myeloid cells 1, which is an immunoglobulin (Ig) superfamily activating receptor that mediates the production of cytokines, chemokines, and ROS. This dysregulated signal transduction may reflect alterations in membrane lipid composition and lipid rafts that result in inappropriate localization or retention in membrane signaling domains. For example, the negative regulator Src homology domain-containing protein tyrosine phosphatase 1 is excluded from lipid rafts after GM-CSF stimulation in neutrophils from young individuals, but it is retained in rafts in neutrophils from older individuals.[35] Altogether, these findings indicate that aging impairs several signaling pathways in neutrophils, including the generation of the respiratory burst and apoptotic pathways, which may result in reduced protection from microbial infection.

2.2.1.2 Natural killer cells

NK cells, which consist of 10%−15% of lymphocytes in peripheral blood, are innate lymphoid cells and take part in the early defense against intracellular cs. They are cytotoxic non-T lymphocytes. According to differential expression of surface markers CD56 and CD16, three NK subsets can be identified. In the CD56^{bright}CD16$^{neg/dim}$ subset, the cells are more immature and secrete cytokines and chemokines, whereas the main NK cell subset CD56^{dim}CD16$^+$ is made of mature NK cells with high cytotoxic capacity after direct contact with tumor or virus-infected target cells.[36] NK cells have the important ability to produce cytokines and chemokines after cell recognition, so they are critical parts of innate immunity during aging and promote cytotoxicity. CD56^{bright}CD16$^{neg/dim}$ NK cells appear diminished in proportion and cytokine/chemokine secretion with aging.[35] In older adults, cytotoxicity is decreased on a per-cell basis, with an expansion of the CD56^{dim}CD16$^+$ cytotoxic NK cell compartment. An increase in the number of mature NK cells with a significant reduction in the immature NK cell subset was observed with advanced age, probably owing to the impaired production of new NK cells upon aging.[35] Thus, the decline in CD56^{bright} NK cells and the increase in the CD56^{dim}CD57$^+$ subset support that the population of NK cells experiences a process of remodeling with a reduction in the output of more immature CD56^{bright} cells and an accumulation of highly differentiated CD56^{dim}CD57$^+$ NK cells.[34] Nonetheless, human NK cells from healthy subjects aged greater than 90 years are still able to secrete chemotactic cytokines such as IL-8, MIP-1α, and CLL5, and efficiently release these chemokines in response to IL-12 and IL-2, although their production remains lower than that observed in young subjects.[36] In addition, aging does not change total NK cell cytotoxicity, probably because of the increased frequencies of mature NK cells.[36]

2.2.1.3 Monocytes and macrophages

Monocytes are the largest type of leukocyte that can differentiate into macrophages and conventional DCs. Monocytes can be classified into three different subsets according to their phenotype: classical (CD14$^+$CD16$^-$), intermediate (CD14$^+$CD16^{dim}), and nonclassical (CD14^{dim}CD16$^+$) monocytes.[36] Aging has not been shown to alter the absolute number and frequency of overall monocytes significantly in humans. However, aging affects the relative distribution of monocyte subsets, with a marked reduction of the classical subset and an increase in the number of inter-mediate and nonclassical monocytes with profound dysregulation in cytokine secretion and phagocytotic capability after Toll-like receptor (TLR) activation of monocytes.[35] In human aged monocytes, a higher level of IL-8 is produced in response to the stimulation of TLR1/2, TLR2/6,

TLR4, or TLR5. This dysregulation appears to be caused by the decline in TLR1 expression with age, and reduced activation of mitogen-activated protein kinase (MAPK) and extracellular signal-regulated kinase 1/2 (ERK1/2) pathways via TLR1/2 in elderly subjects.[36] In contrast, downstream signaling of TLR5 has been shown to increase, which leads to inflammatory responses in the elderly.[36]

Some studies suggest that healthy aging is associated with a significantly increased proportion of total monocytes, without significant changes in the frequency of the three subsets.[36] Monocytes can be characterized into two profiles based on the expression of CD80 and CD163: inflammatory (M1) and antiinflammatory (M2). CD80 is expressed on M1 macrophages, whereas CD163 is expressed on M2. Distinct age-related trends for classical and nonclassical M2 monocytes are shown with a reduction in M2 cells in classical monocytes and an increase in nonclassical monocytes. However, because classical monocytes account for 80%–90% of circulating monocytes, healthy ageing seems to be characterized by a reduced proportion of M2 monocytes.[36] Moreover, stimulation of aged monocytes with TLR4, TLR7/8, and RIG-I agonists produces a lesser amount of cytokines and chemokines such as interferon (IFN)-alfa, IFN-gamma, IL-1β, CCL20, and CCL8, and higher expression of CX3CR1 compared with young monocytes.[36]

Macrophages are important components of innate immunity. They are a type of white blood cell found in the immune system that engulfs and digests pathogens, such as cancerous cells, microbes, cellular debris, and foreign substances that do not have proteins specific to healthy body cells on their surface. The process is called phagocytosis; it acts to defend the host against infection and injury.[34] The macrophage behavior is frequently associated with their state of polarization, which is dichotomized as proinflammatory M1 and antiinflammatory and fibrotic M2. M1/M2 polarization is an intricate product of the interaction between macrophage stimulation, metabolic status, and cellular stress.[34] During aging, endoplasmic reticulum stress, altered nutrient sensing, and inflamed surroundings promote the accumulation of nonconventional, alternatively activated proinflammatory M2 cells. Macrophages also present functional alterations with aging. Aged macrophages derived from different sites (e.g., circulating monocytes, peritoneum, skin, brain) present unchanged or compromised phagocytosis activity as well as altered response to lipopolysaccharides (either higher or lower response, compared with young counterparts).[36] In addition, the macrophages show many age-related functional changes, among which is a reduced expression of the principal TLRs.[35,36] TLRs can recognize pathogen patterns from viruses, bacteria, or fungi, induce nuclear factor (NF)-κB proinflammatory signaling, release different cytokines, and activate innate immunity to eliminate antigens. During aging, TLR-induced IL-6 and TNF-α production are reduced in response to the engagement of TLR1/2 but increased upon TLR4 stimulation.[36]

2.2.1.4 Dendritic cells

DCs are professional antigen-presenting cells (APCs) classified as myeloid DCs (mDCs) or plasmacytoid DCs (pDCs) with different functional activities. mDCs, produce IL-12 and induce T helper (Th) cell type 1 (Th1) and cytotoxic T lymphocyte responses, whereas pDCs produce IFN-alfa/beta in response to bacteria and viruses. These cells are essential links between the innate and adaptive immune systems because they are the major APCs for naive T cells and heavily influence the polarization of CD4+ T cells into Th1, Th2, and other cell patterns.[34,36]

A few studies reveal that the numbers of circulating pDCs and mDCs are reduced in the healthy elderly compared with healthy young subjects, and even more reduced in weakened elderly people. Still, in a large study involving different tissue samples of patients ranging from age 0 to 93 years, DC distribution in different tissues seemed unchanged during aging.[34] Nonetheless, both mDCs and pDCs from elderly individuals show a significant impairment in secreting TNF-α, IL-6, and IL-12 in response to TLR stimulation, probably owing to lower TLR expression. However, basal production of proinflammatory cytokines in the absence of TLR engagement is higher in cells from older individuals, which suggests a dysregulation of cytokine production that may limit further activation through TLR engagement.[34] Among the current explanations for DCs alteration during aging is the excessive NF-κB stimulation engendered by different processes, such as cellular senescence, damage-associated molecular pattern (DAMP) and pathogen-associated molecular pattern stimulation, and changes in local and circulating cytokine levels.[34]

2.2.2 Adaptive immunity

The adaptive immune system is formed from two types of response: the cell-mediated immune response carried out by T lymphocytes and the humoral immune response controlled by activated B lymphocytes and antibodies. After the recognition of antigens by surface receptors, lymphocytes multiply in large quantities (clonal expansion), differentiating into effector and memory cells. The adaptive immune system is intrinsically associated with and tuned by the innate immune system. Therefore, it follows that as the innate system is altered by aging, adaptive immunity is also impaired.

2.2.2.1 B cells

B cells, also known as B lymphocytes, are a subtype of lymphocytes. Early B-cell development and commitment to the B-cell lineage occurs in the fetal liver prenatally, before continuing to mature in the bone marrow throughout life. They function in the humoral immunity component of the adaptive immune system and are responsible for mediating the production of antigen-specific Ig (antibody) directed against invasive pathogens. B cells produce nonsecreted antibody molecules that are inserted into the plasma membrane, where they serve as part of B-cell receptors. When the naive or memory B cells are activated by the antigens, they proliferate and differentiate into antibody-secreting effector cells, known as plasmablasts or plasma cells. In addition, B cells are classified as professional APCs that present antigens and secrete cytokines.

During aging, peripheral B cell changes with declines in cell number and percentage as well as impaired specific humoral immune responses against extracellular pathogens. In particular, B-cell repertoire diversity, Ig isotypes, and receptor repertoire are affected by age.[37,38] In aging, the downregulation of transcriptional factor E47 that controls B-cell functions leads to the reduction of the activation-induced cytidine deaminase that induces class switch recombination and Ig somatic hypermutation, diminished antibody avidity and antibody-mediated protection, and decreased interaction between B cells and CD40L[+] T Th cells.[37,38] Furthermore, elevated levels of TNF-α, typical of inflammaging, can cause human unstimulated B cells from elderly individuals to release significantly higher levels of TNF-α compared with those from young subjects and render them unable to respond to exogenous antigens, mitogens, or vaccines.[36] Besides, the percentage of switched memory B cells decreases while the percentage of late or exhausted memory B cells increases.[36] In addition, the aged bone marrow is less effective in selecting autoreactive B cells for elimination. These processes lead to the accumulation of long-lived age-associated B cells (ABCs), which are identified in both mice and humans. They are refractory to B-cell receptor ligation but are activated by TLR7 or TLR9 signals.[39] ABCs can be generated from naive follicular B cells and possibly other preimmune B cell subsets, such as MZ, B1, and transitional B

214 Molecular, Cellular, and Metabolic Fundamentals of Human Aging

cells. Their differentiation requires TLR9 or TLR7 signals in the context of Th1 cytokines, particularly IFN-gamma and IL-21. In contrast, Th2 cytokines such as IL-4 can impede ABC differentiation.[39] ABCs contribute to immunologic features of aging, including dampened B-cell genesis, altered immune responses to both primary and recall antigen challenges, and an increasing overall inflammatory climate.[39]

2.2.2.2 T cells

Unlike B cells, which are produced and matured in the medullary microenvironment, T cells are produced in the medulla and matured in the thymus. T cells have an important role in orchestrating the immune response and are subdivided into $CD4^+$ and $CD8^+$ T-cell populations with different functions. $CD4^+$ T cells are crucial in achieving a regulated effective immune response to pathogens and possess effector functions. Naive $CD4^+$ T cells can differentiate into one of several lineages of Th cells, including Th1, Th2, and Th17, and Tregs, as defined by their pattern of cytokine production and function. $CD8^+$ T cells constitute an essential branch of adaptive immunity contributing to the clearance of intracellular pathogens and providing long-term protection.

The production of reactive T cells occurs more robustly during childhood, but the maintenance of production occurs throughout life. Two main changes in the adaptive immune system that characterize aging are (1) a decrease in naive T cells that leads to shrinking of the TCR repertoire, and (2) an increase in memory T cells primed by different antigens and upregulation of proinflammatory molecules.[34,40] As described earlier, the involution of the thymus considerably reduces the production of T cells. With the progression of age and evolution of the immunosenescence process, the reduced number of naive T lymphocytes but an increased number of memory T lymphocytes contribute to a greater risk for the occurrence of infectious diseases, cancer, and autoimmune diseases observed with aging.[40] Moreover, the compromised ability of $CD4^+$ T cells to differentiate into functional subsets during aging results in many dysregulated responses including reduced cognate help to B cells with consequent reduced humoral immunity and the increased ratio of the proinflammatory Th17 cells and the immunosuppressive Tregs, favoring a basal proinflammatory status.[36] Thus, changes in the Th17/Treg ratios and altered cytokine expression during aging may contribute to an imbalance between the proinflammatory and antiinflammatory immune response, indicating higher susceptibility to developing inflammatory diseases with increasing age.[36]

Peripheral lymphocyte telomere length is a marker of biological aging and health. T-cell differentiation leads to lower telomerase expression and is associated with a lower T-cell proliferation. In addition, mitochondrial stress is related to oxidative stress, which accelerates telomere shortening in T cells.[40] Although they seem to lack some of the classic markers of cellular senescence, senescent-like T cells acquire a specific phenotype related to such a cellular state, characterized by loss of CD27 and CD28, as well as the expression of CD57, CD45RA, and/or KLRG-1. The senescence of T cells may compromise their senescent-cell elimination response. Mechanistically, senescent cells increase expression of human leukocyte antigen E (HLA-E), whereas mature T cells are characterized by a high expression of natural killer group protein 2A (NKG2A).[36]

3. Interplay between immunosenescence and aging process

3.1 Inflammaging and immunosuppressive network

Inflammaging refers to the state of low-grade, sterile chronic inflammation that develops with age. It is characterized by high serum concentrations of inflammatory cytokines and mediators such as C-reactive protein (CRP), IL-6, IL-8, and TNF. Immunosenescence is one of the biological basic mechanisms that contribute to the onset and progression of inflammaging.[36,41] On the other hand, inflammaging has been attributed to evoking the compensatory immunosuppressive network, and thus it enhances remodeling of the immune system with aging. There is convincing evidence that the presence of immunosuppressive cells increases with aging.[42] For instance, with aging, the numbers and immunosuppressive activity of myeloid-derived suppressor cells (MDSCs) increase in the circulation, bone marrow, spleen, and LN of humans and mice.[36] In addition, there is a significant increase in the numbers of Tregs in the circulation, skin, and adipose tissue with aging. Interestingly, senescent stromal cells in mouse skin induce a local inflammation that recruits immunosuppressive MDSCs and Tregs into senescent skin. Senescent skin cells contain an increased level of proinflammatory factors (e.g., IL-6, CCL2, CXCL1, and GM-CSF) as well as immunosuppressive markers IL-10, TGF-β, and arginine 1.[42] Many studies have revealed a significant age-related increase in the numbers of M2 macrophages in several tissues, including mouse bone marrow, spleen, lungs, and skeletal muscles. For instance, M2 macrophages from geriatric mice display robust immunosuppressive activity by secreting an increased level of TGF-β and IL-10 cytokines compared with their young counterparts.[42] Inflammaging affects the hematopoietic system, inducing the myeloid-biased shift toward myelopoiesis. Inflammatory factors such as colony-stimulating factors (CSFs) and interferons regulate the expansion of the myelopoietic lineage and can trigger the formation of MDSCs from the myelopoietic pathway.[43] MDSCs are immature myeloid cells that can evoke the differentiation of immunosuppressive Tregs from effector T cells. Inflammatory mediators recruit proinflammatory cells as well as immunosuppressive MDSCs and Tregs into inflamed tissues.[42,43] In addition, inflammaging leads to the accumulation of other immunosuppressive phenotypes of immune cells such as Tregs, B cells, DC cells, NK cells, NKT cells (type II NKT), and macrophages (M2 phenotype).[42]

3.2 Feed-forward regulation between cellular senescence and immunosuppression

Feed-forward regulation mechanism exists between cellular senescence and immunosuppression, which enhances the appearance of a chronic inflammatory state and compensatory immunosuppression.[43] The accumulation of senescent cells in aging tissues triggers the secretion of inflammatory mediators such as CSFs, cytokines, and chemokines, leading to myelopoiesis and subsequently the recruitment of immune cells into the affected tissues. Immunosuppressive cells secrete ROS and reactive nitrogen species (RNS) and antiinflammatory cytokines such as TGF-β, IL-4, and IL-10, which suppress the function of NK and CD8$^+$ T cells and thus inhibit the clearance of senescent cells.[42,43] In addition, senescent cells can enhance their accumulation within aging tissues. The robust expression of cell-surface NKG2D ligands and their soluble cleavage products as well as the expression of surveillance inhibitors such as HLA-E and CEACAM1 prevent the clearance of senescent cells by NK and CD8$^+$ T cells.[43] In addition to increasing the numbers of senescent cells in aging tissues, this feed-forward circuit impairs tissue homeostasis and promotes tissue degeneration through the persistent occurrence of immunosuppressive cells with aging.[43] For example, some immunosuppressive cytokines, especially TGF-β, induce disturbances in the structures of the ECM by increasing the secretion of proteolytic enzymes and enhancing fibrosis in many tissues. Moreover, senescent cells themselves can modify the structures of tissue ECM through their secretomes. Secreted proteinases, such as matrix metalloproteinases, collagenases, and elastases, produce many danger-associated molecular patterns from the ECM-derived DAMPs.[43] These observations indicate that cellular senescence, inflammaging, and compensatory immunosuppression are partners in the feed-forward circuit that progressively promotes the aging process in tissues.

3.3 Immunosenescence and age-associated diseases

The expansion of late-differentiated T cells (CD28$^-$) increases serum levels of autoantibodies and proinflammatory cytokines that are implicated in morbidities during aging. Features of accelerated immunosenescence can be identified in adults with chronic inflammatory conditions and are predictive of poor clinical outcomes. Therefore, an interplay exists between immunosenescence and age-related diseases such as neurodegenerative diseases, rheumatoid arthritis (RA), cancer, cardiovascular, and metabolic diseases.[44]

3.3.1 Neurodegenerative diseases

Immunosenescence has been repeatedly implicated in cognitive processes and neurodegenerative diseases. The most common age-related neurodegenerative diseases include Alzheimer's disease (AD) and Parkinson's disease (PD), and inflammaging has been implicated in cognitive decline and dementia during aging.[44] Peripheral inflammatory mediators, especially CRP, have been connected with cognitive impairment in the elderly. Also, the levels of plasma CRP can be used to predict future memory impairments.[35] Besides, the presence of CRP is related to the formation of β-amyloid plaques in AD, and individuals with PD whose mental faculties are reduced exhibit increased plasma levels of CRP.[45] In addition, high inflammation (CRP and IL-6 levels) and metabolic syndrome collectively exacerbate cognitive impairment. The excessive and/or prolonged release of proinflammatory cytokines (TNF-α, IL-1β, IL-6, IL-10, and IL-12) in the central nervous system results in reduced brain-derived neurotrophic factor levels, glutamatergic activation (excitotoxicity), oxidative stress, and induction of apoptosis, which contribute to cognitive decline.[43,44] Neuroinflammation is known to weaken various brain functions, such as inhibition of hippocampal neurogenesis, and to impair cognitive abilities. However, under physiologic conditions, proinflammatory cytokines are beneficial for providing trophic support to neurons and strengthening neurogenesis, contributing to cognitive function.[44]

T-cell senescence has also been implicated in cognitive decline.[44] Different stages of T-cell differentiation can be determined based on the cell-surface expression of costimulatory molecules CD27 and CD28, namely early differentiated (CD27$^+$CD28$^+$), intermediate-differentiated

(CD27⁻CD28⁺), and late-differentiated or aged T cells (CD27⁻CD28⁻). CD28⁻ T cells have shortened telomeres and display differential features (cytotoxic, immunosuppressive, or regulatory) under various conditions. In peripheral blood of AD patients, decreased percentages of naive T cells, elevated memory cells, and great expansion in late differentiated CD28⁻ T cells (in both CD4⁺ and CD8⁺ populations) may all be seen compared with healthy young or older adults.[36,45] Under nonpathologic conditions, better cognitive performance is associated with lower numbers of effector memory CD4⁺ T cells and higher numbers of naive CD8⁺ T and B cells. In addition, increased numbers of CD8⁺ T effector memory CD45RA⁺ cells (TEMRA) are found in peripheral blood mononuclear cells and in the brain adjacent to Aβ plaques of AD patients. There, cells are negatively associated with cognition.[45]

Moreover, dysregulated innate immune responses in the aging brain may be relevant to AD. High expression of TLR1−9 is found in both mouse and human microglia. Widespread TLR expression is also found in mouse astrocytes and cortical neurons, although in humans only TLR3 expression has been reported.[35] In humans, there is marked upregulation of TLR- and inflammasome-associated genes in older adults compared with young individuals. The effect of TLR activation in the context of neurodegeneration has been investigated mainly in mouse models and seems to be complex.[35] The Aβ peptide that is found in amyloid plaques deposited in the brains of individuals with AD induces a TLR-dependent innate immune response that may facilitate Aβ clearance.[35,45] However, systemic inflammation, such as by administration of lipopolysaccharide, exacerbates Aβ plaque formation and cognitive impairment in a transgenic AD model, and an association between systemic inflammation and AD was reported in humans. In addition, the NLRP3 inflammasome is implicated in AD pathogenesis.[35,45]

3.3.2 Rheumatoid arthritis

Aging can be associated with an increased risk for autoimmunity, which can be central or peripheral. Thymic involution, in particular the diminution in numbers of medullary TECs, which have an essential role in establishing central T-cell self-tolerance, may increase the potential risk for the breakdown of self-tolerance. On the other hand, homeostatic proliferation (HP) of naive T cells in the periphery also favors the development of autoimmunity, partially because persistent HP may cause a biased increase in the T-cell population bearing higher intrinsic affinity to self−major histocompatibility complex molecules.[44] RA is a chronic inflammatory autoimmune disease associated with symmetrical and destructive inflammation in joints and other tissues. Several premature immunosenescent features, including decreased thymic functionality, expansion of late-differentiated effector T cells, increased telomeric attrition, and increased production of proinflammatory cytokines, have been observed in RA.[44] Specifically, a significant expansion of late differentiated T cells ($CD4^+CD28^-$ and $CD8^+CD28^-$) in RA, which terminally differentiate into TEMRA cells, leads to inflammaging and further aggravating RA through increased production of TNF-α, IL-1β, IL-6, and IFN-gamma under stimulation.[40] The expansion of late-differentiated or aged $CD28^-$ T cells is linked to disease severity in RA. Furthermore, RA is associated with the accumulation of other senescent cells, including senescent synovial fibroblasts ($P16INK4a^+$), which display an enhanced inflammatory phenotype.[40,44]

3.3.3 Cancer

The incidence of cancer increases with age. It is known that the immune system can recognize and destroy the precursors of cancer, which is largely mediated by cytotoxic $CD8^+$ T and NK cells.[46] As described earlier, age-related thymic involution leads to the reduced output and function of T cells, which contribute to tumor development and progression.[40,46] Aging also leads to the accumulation of cells with suppressor actions, characterized by the increased frequencies of Tregs and MDSCs. These cells inhibit the generation of antitumor responses by several mechanisms such as the secretion of cytokines (e.g., IL-10, TGF-β).[44] In addition, the number of senescent tissue cells increases during aging, which might exaggerate the inflammatory environment, favoring cancer development through the initiation, promotion, and progression of tumors.[40,46] In terms of the tumor microenvironment, the aged microenvironment has a key role in reprogramming tumor cells toward a senescence-associated secretory phenotype (SASP).[40,46] Although several chemotherapeutic strategies induce senescence in cancer, SASP can contribute negatively to cancer therapy, such as the increase in malignant phenotypes and tumor induction.[44]

3.3.4 Cardiovascular diseases

Aging is the most important determinant of cardiovascular diseases (CVDs). CVDs are a group of heart and blood vessel disorders connected to features of organismal aging and loss of homeostasis. They contribute to increased morbidity and mortality rates. Elevated levels of proinflammatory cytokines contribute to the inflammaging process, augmenting the probability of endothelial damage, impairment of vascular remodeling, and the development of atherosclerosis and insulin resistance. During cardiac stress, ischemic injury, hypertension, and metabolic syndrome, necrotic cells release high amounts of high-mobility group box 1 and heat shock protein 60.[44] These DAMPs are recognized by pattern recognition receptors expressed mainly by innate immune cells.[35,44] As a result, tissue cells (mainly M1 macrophages) and nonimmune cells secrete large amounts of proinflammatory cytokines to recruit phagocytic immune cells, clear apoptotic and pyroptotic cells, and renew tissue. Classical biomarkers of inflammation of the innate immune response, including CRP, IL-1β, IL-6, TNF-α, and several cell adhesion molecules, are similarly linked to the occurrence of myocardial infarction and stroke in both healthy individuals and those known to have coronary diseases.[44] Elevated secretion of TNF-α, IL-6, and IL-1β from cardiomyocytes and peripheral tissues have been shown to have an important role in the pathogenesis and progression of myocardial dysfunction, and plasma levels of these proinflammatory cytokines can predict short- and long-term survival in patients with heart failure.[44]

Proinflammatory M1 macrophages, inflammatory cytokines (such as IL-6 and IL-12), ROS, and RNS that are present in atherosclerotic plaques have been implicated in acute coronary syndrome (ACS). In addition, increased levels of Th1 and Th17 and its cytokines, such as IL-17, IL-21, and IL-23, have been described in atherosclerotic carotid artery plaques.[44] Intermediate senescent CD14$^+$CD16$^+$ monocytes are also present in the atherosclerotic plaques and express high levels of the vascular adhesion molecules necessary for adhesion and diapedesis through endothelial cells.[35] In addition to the innate immune system, senescent T cells are responsible for CVD pathogenesis. In atherosclerosis and ACS, CD8$^+$CD28$^-$ T-cell populations are expanded and constitute a risk factor for vascular dysfunction in a cohort of cytomegalovirus–infected individuals. Peripheral late-differentiated CD4$^+$CD28$^-$ T cells are present in acute coronary events, and CD4$^+$ effector memory T cells (CD3$^+$CD4$^+$CD45RA$^-$CD45RO$^+$CCR7$^-$) are associated with atherosclerosis.[40,46]

3.3.5 Metabolic diseases

The systemic low-grade chronic inflammatory process present in elderly individuals is a critical etiologic component of physiologic decline and a risk factor for age-related diseases. Metabolically driven inflammation, termed metainflammation, is a major aspect of metabolic disorders that resemble the critical inflammatory process of aging. Metabolic diseases such as type 2 diabetes (T2D) are mainly attributed to metainflammation.[47] T2D widely displays inflammaging and metainflammation. One of the first mechanisms described to promote insulin resistance is the interference of insulin signaling driven by TNF-α, establishing the first connection between metabolism and immune response.[44] Since this finding, several studies have been performed to clarify the link between T2D and immunity. For example, a classical feature of immunosenescence presented in T2D is the decreased pool of CD4[+]-naive T cells, concomitant with an increased pool of memory CD4[+] T cells and effector CD4[+] and CD8[+] T cells.[44] The heightened population of effector T cells has been identified as the major producer of IFN-gamma and TNF-α, enhancing the systemic proinflammatory status. Late-differentiated or senescent T cells (CD8[+]CD57[+] and CD8[+]CD28[−]) have been demonstrated to predict the development of hyperglycemia in humans.[44] T2D patients have a reduced T-cell repertoire related to the increased naive/effector T-cell ratio. However, both adaptive and innate immune compartments show impaired function and activation in T2D. Diminished phagocytic activity and TLR responsiveness are found in peripheral blood monocytes of T2D patients.[44] In addition, the heightened proinflammatory B-cell pool is implicated as the main driver of T-cell inflammatory profile in T2D.[48] Inflammatory B cells are characterized by increased basal secretion of IL-6 and IFN-gamma concomitant with accentuated decreased secretion of IL-10, sustaining chronic low-grade inflammation.[48]

4. Conclusions

When kept under a certain threshold, chronic inflammatory stimulation should not be considered detrimental, because it pushes a secondary adaptive activation of antiinflammatory networks. The strength of the maladaptive response is likely critical to determine different aging trajectories and the net outcome: unsuccessful aging and age-associated diseases rather than successful aging and longevity.

The antiinflammatory response represents a dynamic and active process able to trigger specific molecular pathways aimed at inhibiting and resolving dangerous inflammation. Consistent with this, the development of age-related diseases and weakness is a result of excessive stimulation of proinflammatory responses but also an ineffective antiinflammatory reaction, whereas the attainment of longevity and successful aging is determined by a reduced predisposition to stimulate inflammatory pathways in addition to an effective antiinflammatory response. In other words, individuals who have well-preserved and organized antiinflammatory activity are able to counteract the age-related increase in inflammatory markers (inflammaging), and the probability of developing age-related diseases is highly reduced or delayed, or shows less severe consequences.

Centenarians who represent the best example of successful aging have a large quantity of circulating antiinflammatory molecules such as TGF-β1, IL-10, IL-1 receptor antagonist (IL-1Ra), adiponectin, cortisol, antiinflammatory arachidonic acid compounds, including hydroxyeicosatetraenoic acid (HETE) and epoxyeicosatrienoic acid (EET), and mitokines, such as fibroblast growth factor 21 (FGF21), growth differentiation factor 15 (GDF15), and humanin (HN). However, this antiinflammatory state is effectively triggered to counterbalance the concomitant increased levels of inflammatory molecules in plasma, such as IL-6, IL-15, IL-18, IL-18 binding protein, IL-22, CRP, serum-amyloid A, fibrinogen, von Willebrand factor, resistin, and leukotrienes. It remains unknown whether this optimal balance is a characteristic of these individuals during their entire lifetime as a result of both lifestyle and genetic background or whether they acquire this ability in the later phase of life owing to an adaptive strategy.

References

1. Geiger H, de Haan G, Florian MC. The ageing haematopoietic stem cell compartment. *Nat Rev Immunol.* 2013;13:376–389.
2. Rossi DJ, Bryder D, Zahn JM, et al. Cell intrinsic alterations underlie hematopoietic stem cell aging. *Proc Natl Acad Sci U S A.* 2005;102:9194–9199.
3. Beerman I, Maloney WJ, Weissmann IL, Rossi DJ. Stem cells and the aging hematopoietic system. *Curr Opin Immunol.* 2010;22:500–506.
4. Beerman I, Bhattacharya D, Zandi S, et al. Functionally distinct hematopoietic stem cells modulate hematopoietic lineage potential during aging by a mechanism of clonal expansion. *Proc Natl Acad Sci U S A.* 2010;107:5465–5470.
5. Sudo K, Ema H, Morita Y, Nakauchi H. Age-associated characteristics of murine hematopoietic stem cells. *J Exp Med.* 2000;192:1273–1280.
6. Mercier FE, Ragu C, Scadden DT. The bone marrow at the crossroads of blood and immunity. *Nat Rev Immunol.* 2011;12:49–60.
7. Signer RA, Montecino-Rodriguez E, Witte ON, McLaughlin J, Dorshkind K. Age-related defects in B lymphopoiesis underlie the myeloid dominance of adult leukemia. *Blood.* 2007;110:1831–1839.
8. Muller-Sieburg CE, Sieburg HB. Clonal diversity of the stem cell compartment. *Curr Opin Hematol.* 2006;13:243–248.
9. Dykstra B, Olthof S, Schreuder J, Ritsema M, de Haan G. Clonal analysis reveals multiple functional defects of aged murine hematopoietic stem cells. *J Exp Med.* 2011;208:2691–2703.

Aging and the immune system **223**

10. Ergen AV, Boles NC, Goodell MA. Rantes/Ccl5 influences hematopoietic stem cell subtypes and causes myeloid skewing. *Blood*. 2012;119:2500–2509.

11. Tuljapurkar SR, McGuire TR, Brusnahan SK, et al. Changes in human bone marrow fat content associated with changes in hematopoietic stem cell numbers and cytokine levels with aging. *J Anat*. 2011;219:574–581.

12. Kohler A, Schmithorst V, Filippi MD, et al. Altered cellular dynamics and endosteal location of aged early hematopoietic progenitor cells revealed by time-lapse intravital imaging in long bones. *Blood*. 2009;114:290–298.

13. Katayama Y, Battista M, Kao WM, et al. Signals from the sympathetic nervous system regulate hematopoietic stem cell egress from bone marrow. *Cell*. 2006;124:407–421.

14. Chaudhry MS, Velardi E, Dudakov JA, van den Brink MR. Thymus: the next (re) generation. *Immunol Rev*. 2016;271:56–71.

15. Elyahu Y, Monsonego A. Thymus involution sets the clock of the aging T-cell landscape: implications for declined immunity and tissue repair. *Ageing Res Rev*. 2021;65:101231.

16. Park JE, Botting RA, Dominguez Conde C, et al. A cell atlas of human thymic development defines T cell repertoire formation. *Science*. 2020;367.

17. Akbar AN, Henson SM. Are senescence and exhaustion intertwined or unrelated processes that compromise immunity? *Nat Rev Immunol*. 2011;11:289–295.

18. Mo R, Chen J, Han Y, et al. T cell chemokine receptor expression in aging. *J Immunol*. 2003;170:895–904.

19. Dudakov JA, Goldberg GL, Reiseger JJ, Vlahos K, Chidgey AP, Boyd RL. Sex steroid ablation enhances hematopoietic recovery following cytotoxic antineoplastic therapy in aged mice. *J Immunol*. 2009;183:7084–7094.

20. Taves MD, Mittelstadt PR, Presman DM, Hager GL, Ashwell JD. Single-cell resolution and quantitation of targeted glucocorticoid delivery in the thymus. *Cell Rep*. 2019;26:3629–3642 e3624.

21. Bronte V, Pittet MJ. The spleen in local and systemic regulation of immunity. *Immunity*. 2013;39:806–818.

22. Turner VM, Mabbott NA. Influence of ageing on the microarchitecture of the spleen and lymph nodes. *Biogerontology*. 2017;18:723–738.

23. Budamagunta V, Foster TC, Zhou D. Cellular senescence in lymphoid organs and immunosenescence. *Aging (Albany NY)*. 2021;13:19920–19941.

24. Bajenoff M, Egen JG, Qi H, Huang AY, Castellino F, Germain RN. Highways, byways and breadcrumbs: directing lymphocyte traffic in the lymph node. *Trends Immunol*. 2007;28:346–352.

25. Thompson HL, Smithey MJ, Surh CD, Nikolich-Zugich J. Functional and homeostatic impact of age-related changes in lymph node stroma. *Front Immunol*. 2017;8:706.

26. Cakala-Jakimowicz M, Kolodziej-Wojnar P, Puzianowska-Kuznicka M. Aging-related cellular, structural and functional changes in the lymph nodes: a significant component of immunosenescence? An overview. *Cells*. 2021;10.

27. Zolla V, Nizamutdinova IT, Scharf B, et al. Aging-related anatomical and biochemical changes in lymphatic collectors impair lymph transport, fluid homeostasis, and pathogen clearance. *Aging Cell*. 2015;14:582–594.

28. Becklund BR, Purton JF, Ramsey C, et al. The aged lymphoid tissue environment fails to support naive T cell homeostasis. *Sci Rep*. 2016;6:30842.

29. Richner JM, Gmyrek GB, Govero J, et al. Age-dependent cell trafficking defects in draining lymph nodes impair adaptive immunity and control of west nile virus infection. *PLoS Pathog*. 2015;11:e1005027.

30. Textor J, Mandl JN, de Boer RJ. The reticular cell network: a robust backbone for immune responses. *PLoS Biol*. 2016;14:e2000827.

31. Masters AR, Hall A, Bartley JM, et al. Assessment of lymph node stromal cells as an underlying factor in age-related immune impairment. *J Gerontol A Biol Sci Med Sci*. 2019;74:1734–1743.

32. Thannickal VJ, Zhou Y, Gaggar A, Duncan SR. Fibrosis: ultimate and proximate causes. *J Clin Invest.* 2014;124:4673−4677.
33. Thompson HL, Smithey MJ, Uhrlaub JL, et al. Lymph nodes as barriers to T-cell rejuvenation in aging mice and nonhuman primates. *Aging Cell.* 2019;18:e12865.
34. Rodrigues LP, Teixeira VR, Alencar-Silva T, et al. Hallmarks of aging and immuno-senescence: connecting the dots. *Cytokine Growth Factor Rev.* 2021;59:9−21.
35. Shaw AC, Goldstein DR, Montgomery RR. Age-dependent dysregulation of innate immunity. *Nat Rev Immunol.* 2013;13:875−887.
36. Santoro A, Bientinesi E, Monti D. Immunosenescence and inflammaging in the aging process: age-related diseases or longevity? *Ageing Res Rev.* 2021;71:101422.
37. Frasca D, Diaz A, Romero M, Garcia D, Blomberg BB. B cell immunosenescence. *Annu Rev Cell Dev Biol.* 2020;36:551−574.
38. Bulati M, Buffa S, Candore G, et al. B cells and immunosenescence: a focus on IgG+IgD-CD27- (DN) B cells in aged humans. *Ageing Res Rev.* 2011;10:274−284.
39. Cancro MP. Age-associated B cells. *Annu Rev Immunol.* 2020;38:315−340.
40. Mittelbrunn M, Kroemer G. Hallmarks of T cell aging. *Nat Immunol.* 2021;22:687−698.
41. Franceschi C, Bonafe M, Valensin S, et al. Inflamm-aging. An evolutionary perspective on immunosenescence. *Ann N Y Acad Sci.* 2000;908:244−254.
42. Salminen A. Activation of immunosuppressive network in the aging process. *Ageing Res Rev.* 2020;57:100998.
43. Salminen A. Feed-forward regulation between cellular senescence and immunosup-pression promotes the aging process and age-related diseases. *Ageing Res Rev.* 2021;67:101280.
44. Barbe-Tuana F, Funchal G, Schmitz CRR, Maurmann RM, Bauer ME. The interplay between immunosenescence and age-related diseases. *Semin Immunopathol.* 2020;42:545−557.
45. Wu KM, Zhang YR, Huang YY, Dong Q, Tan L, Yu JT. The role of the immune system in Alzheimer's disease. *Ageing Res Rev.* 2021;70:101409.
46. Minato N, Hattori M, Hamazaki Y. Physiology and pathology of T-cell aging. *Int Immunol.* 2020;32:223−231.
47. Hotamisligil GS. Inflammation, metaflammation and immunometabolic disorders. *Nature.* 2017;542:177−185.
48. DeFuria J, Belkina AC, Jagannathan-Bogdan M, et al. B cells promote inflammation in obesity and type 2 diabetes through regulation of T-cell function and an inflammatory cytokine profile. *Proc Natl Acad Sci U S A.* 2013;110:5133−5138.

CHAPTER 8

Canonical and novel strategies to delay or reverse aging

Brian C. Gilmour[1,2], Linda Hildegard Bergersen[1,3,4,5] and Evandro Fei Fang[1,6]

[1]The Norwegian Centre on Healthy Ageing (NO-Age) Network, Oslo, Norway; [2]Department of Molecular Medicine, Institute of Basic Medical Sciences, University of Oslo, Oslo, Norway; [3]The Brain and Muscle Energy Group, Electron Microscopy Laboratory, Department of Oral Biology, University of Oslo, Oslo, Norway; [4]Synaptic Neurochemistry and Amino Acid Transporters Labs, Division of Anatomy, Department of Molecular Medicine, Institute of Basic Medical Sciences and Healthy Brain Ageing Centre, University of Oslo, Oslo, Norway; [5]Center for Healthy Ageing, Department of Neuroscience and Pharmacology, Faculty of Health Sciences, University of Copenhagen, Copenhagen, Denmark; [6]Department of Clinical Molecular Biology, University of Oslo and Akershus University Hospital, Lørenskog, Norway

1. Approaching aging as a treatable disease

The possibility of treating aging in the same way as other diseases, and as a general prophylactic treatment for diseases that occur in old age, is a wide-ranging question with broad societal and economic impacts, and with manifold dimensions. For example, ethical dilemmas arise if the treatment of aging is limited to those who can afford to pay premiums of care. Likewise, the increase in working age that could follow an increase in the health span of populations could vastly change the workforce balance, resulting in large-scale economic repercussions if not considered.

Such a multifaceted problem must be carefully investigated as it develops, with rigorous attention to each facet. However, to keep the scope more manageable, this chapter is limited to current understandings, ways of life, and burgeoning treatments that have arisen from early studies into and related to aging, while lightly touching on newer, emerging sectors that may occupy some larger role in any pharmaceutical treatment of aging.

If we look at aging as a disease, with specific faults that leads to a larger syndrome, reducing, halting the fault, or repairing its detrimental effects can be used to treat it.

The unique generation of aging and diseases induced or exacerbated by the aging process makes treatments that work well for more classical diseases less viable. Such classical diseases are defined by an invading pathogen: a bacteria, virus, parasite, or other such foreign organism that attempts to compete for resources available to the body. All such pathogens are

Molecular, Cellular, and Metabolic Fundamentals of Human Aging
ISBN 978-0-323-91617-2
https://doi.org/10.1016/B978-0-323-91617-2.00005-5

© 2023 Elsevier Inc.
All rights reserved.

characterized by being foreign to the body, and thus possess unique markers and signatures that can be used by the body's immune system to find and eliminate the invading pathogen. This is not as easy in the case of aging, in which targets to be removed are the body's own cells, portions of larger tissues, which make it harder to distinguish between them and their healthier counterparts.

Thus, difficulties of treating aging resemble those often encountered in treating cancer, which also originates from the body's own tissues, and although this difficulty has long hampered progress in treating both conditions, the aging field should perhaps look with hope to the current expansion and diversification of therapies, immunologic and otherwise, occurring in the field of cancer therapy.

Although aging is a different beast from cancer, and has peculiarities that must be considered, a selection of mechanisms have significant effects on both the life span and the occurrence of cancer, such as the control and regulation of the metabolism, and the control and quality of DNA repair. The outsized importance of these other mechanisms makes them invaluable in the control of aging, even on a prepharmaceutical level.

2. Nonpharmaceutical interventions against aging

Pharmacy against aging is still a nascent, if active field, but research has long been carried out on the underlying mechanisms of aging, and it was discovered that certain habits and lifestyles predispose to better overall health in advanced age. This next section will discuss several examples of these habits and delve into what is understood about their mechanism of action with regard to aging.

Many folk treatments, habits, diets, and exercises are touted as ways to reduce the detrimental effects of aging. We have all heard that remaining active during old age is crucial for retaining and improving physical fitness; that stimulating the mind by learning and puzzle-solving is crucial to keeping our thoughts sharp, and that eating a balanced diet can have a surprising effect on health in old age.

Such folk treatments abound in the common treatment of all diseases. As in many fields, there are many fringe beliefs with limited physical backing, but likewise there are certain beliefs that spring not from desire, superstition, or wish, but rather from observation, and observation is the backbone of all science. In addition, many of these observations have helped to unravel fundamental concepts that drive the aging process, producing druggable targets and pathways that continue to fuel the expanding field of antiaging therapies. Thus, before we talk about pharmacy in aging, we will begin with some of the more nonpharmaceutical treatments and what is known about their methods of action.

2.1 Calorie restriction and fasting

As mentioned, metabolism is thought to have a significant effect on both cancer prognosis and aging. However, many parts of human metabolism are not influenceable without pharmaceutical intervention. Metabolic rates, for example, are largely fixed for each person, and are altered at key life milestones rather than generally with age, as was thought until recently.[1]

Thus, in terms of metabolism, diet is the factor most under individual control. However, the ability for dietary changes to influence life span was demonstrated as early as 1935, when Clive McCay and Mary F. Crowell [2] noticed that a calorie-restricted diet (CR) increased life span in mice compared with mice allowed to eat at will. Similar results have since been demonstrated in *Caenorhabditis elegans* and *Drosophila melanogaster*, and in primates, with some life spans extended more than 50%.[3-6]

Although the data from such methods have withstood scrutiny well, there have been difficulties translating such discoveries to clinical or other applications in the general population. Compliance with the strict caloric requirements needed for the beneficial effects of CR in humans is low and may be so limited as to induce a near-anorexic state.[7]

Thus, other routes have been explored to benefit from these data. Variations on the theme of the CR diet, such as intermittent fasting (IF), have been explored, as well as other fasting-mimicking diets, such as the ketogenic diet.[8] In both of these equivalents, fasting or reduction of caloric intake for an extended period (24–48 h) is compensated by a period of regular caloric intake. Ideally this would provide the benefits of the CR diet without the risk for deleterious caloric restriction.[9]

CR and IF share a common mechanism in that they allow switching between different metabolic profiles (i.e., from cellular repair and recycling to produce needed nutrients during CR and IF to growth, cellular reproduction, and proliferation during periods of caloric intake).[10]

There is less history supporting the beneficial effects of IF, but it has been shown to extend life span in rodent models.[11] In addition, both IF and CR have beneficial effects on general health, such as increasing insulin sensitivity, reducing body weight, and improving cardiometabolic health and lipid profiles. These benefits have been demonstrated in humans for both CR and IF.[12–14] In contrast, other fast-mimicking diets, such as the ketogenic diet, exhibit similar health benefits, but they lack consistent data demonstrating that they increase overall longevity as well.[15,16]

The shared increase in longevity seen with both CR and IF attracted attention to their underlying mechanisms, which has since put the signaling of the mammalian target of rapamycin (mTOR) complexes 1 and 2 (mTORC1/2) in the spotlight. mTOR signaling is one of several systems that sense nutrient availability. Thus, it acts as a key controller of biogenesis, translation, aerobic glycolysis, and general proliferation.[17] Since then, direct inhibition of mTORC1 by genetic ablation or pharmaceutical inhibition through, for example, rapamycin, the drug for which the complex is named, has been shown to increase the life span in numerous species.[18–22] These effects have been demonstrated to result from the direct inhibition of mTORC1 and mTOR signaling.[23–25]

3. Pharmaceutical treatments against aging

3.1 Rapamycin and rapalogs

The case of rapamycin and the potent effects of mTOR inhibition in prolonging the life span and alleviating several aging-related symptoms has spurred investigation into rapamycin analogs, or rapalogs, for use in treating common declines in health seen in human aging.

Whereas rapamycin itself is used clinically to reduce transplant rejection,[26] the use of rapamycin over extended periods is limited by toxic effects. Long-term inhibition of mTOR signaling is detrimental for general cell health, as well as the health of the immune system.[27]

The specific effects of rapamycin on the immune system[28] have been of special interest in applying mTOR inhibition to treat human aging. Immune function declines significantly with age, making the elderly more prone to opportunistic infections and decreasing their responses to vaccines.[29] To treat this decreased vaccine response, elderly patients given a promising rapalog, everolimus (formerly RAD001, Novartis), showed a general increase in antibody titers, which suggested stronger antibody responses to the vaccine in question.[30] This was later backed up by another clinical trial that showed a general decrease in infection rate among elderly people treated with a combination of everolimus and a combined PI3K/mTOR inhibitor, dactolisib (formerly BEZ235, Novartis).[31] More recently, everolimus has found a home as part of combined treatment against cancer.[32,33]

Canonical and novel strategies to delay or reverse aging   **229**

What makes a good rapalog is still not entirely clear, however, as evidenced by a more recent clinical trial of dactolisib (ClinicalTrials: NCT04139915, under the name RTB101), which was withdrawn when it failed to reach its primary end point.

3.2 Mitophagy inducers

Unsurprisingly, given the significant effect of diet alteration and calorie restriction in reducing the detrimental effects of aging, many currently approved treatments against aging take the form of dietary supplements that work by influencing energetics, promoting proper waste disposal, better cellular health, and general cleaning. This section will focus on mitophagy inducers, specifically oxidized nicotine adenine dinucleotide (NAD^+), urolithin A, and actinonin.

Mitophagy is a key part of the cellular recycling system, related to autophagy. Mitophagy controls mitochondrial quality by breaking down and recycling damaged mitochondria.[34] Defects in mitophagy have been linked to aging in general, as well as to several comorbidities of aging and age-predisposed diseases, such as neurodegeneration.[35,36]

3.2.1 Oxidized nicotine adenine dinucleotide

NAD^+ is an interesting molecule in that it is a naturally occurring coenzyme participating in several metabolic pathways. The importance of NAD^+ and the $NAD^+/NADH$ ratio in general health and in aging has been dealt with several times in this book. Apart from its metabolic functions, NAD^+ is crucial for the functions of sirtuins and poly-ADP-ribose polymerases (PARPs).[37–39] NAD^+ levels are known to decrease during aging, affecting the proper function of sirtuins and PARPs.[40–42] In addition to participation in metabolism and the actions of sirtuins and PARPs, low NAD^+ levels have been linked to a decrease in mitophagy and a resulting decrease in mitochondrial quality in a model of Werner syndrome. Supplementation with NAD^+ precursors was sufficient to improve metabolism and mitochondrial quality and increase mitophagy.[43]

NAD^+ itself is impossible to provide directly to cells, but several NAD^+ synthesis pathways exist to produce NAD^+ from circulating precursors, the most common of which are nicotinamide riboside (NR) and nicotinamide mononucleotide (NMN). Given as a supplement, both molecules had positive effects in aging studies in yeast,[44] worms,[45] and mice.[37–39] Although only NR extends the life span in all three species,[46] NMN provides several health benefits in mice without extending their life span.[47,48]

Enough data have been gathered to push both molecules forward into clinical trials. NR has been shown to be readily bioavailable[49] and well-tolerated[50] and to increase NAD^+ levels.[51,52] Clinical trials for NMN are ongoing. NAD^+ precursors are available on the private market as a dietary supplement.

3.2.2 Urolithin A

Urolithin A is a compound produced in the gut from complex polyphenols found in food such as pomegranates, red berries, and walnuts.[53–56] Although the precursors of urolithin A are abundant in food, only 40% of individuals are thought to convert these precursors to active urolithin A. The ability to catalyze urolithin A production is linked to gut health,[57] diet, and the more general health of the individual in question.[58]

When included as a regular dietary supplement, urolithin A was shown to extend the life span and improve fitness of *C. elegans*. The latter was also demonstrated in rodent models of age-related muscular decline.[59] Similar antiaging effects were noted upon treatment of senescent human skin fibroblasts, reducing the expression of several cellular markers of aging, including matrix metalloproteinase-1 (MMP-1) and reactive oxygen species (ROS) levels, although treatment was not enough to reverse the senescent state of the cells.[60] More recently, the promise of urolithin A was demonstrated in a proof-of-concept randomized trial (ClinicalTrials: NCT03464500) in middle-aged adults, in whom treatment produced various improvements in physical fitness, including muscle strength, aerobic endurance, and physical performance. All improvements were linked to urolithin A's effect on mitophagy.[61]

3.2.3 Actinonin

Actinonin differs from other known mitophagy inducers in that it is a naturally occurring inhibitor of the protein peptide deformylase.[62] It is essential for prokaryotes but absent in mammals, and thus functions as an antibacterial molecule. Regardless, actinonin has been shown to increase mitophagy in neuronal stem cells from mice.[63] However, the mechanism behind this increase is still poorly understood, although it may be related to mitochondrial-specific ribosome and RNA depletion.[64] Lack of knowledge about the method of action of actinonin has impeded its clinical development in a way similar to the cases of NAD^+ and urolithin A.

4. Novel approaches to slow or reverse aging

4.1 Cellular reprogramming

The ability to induce cells back to a pluripotent state has been demonstrated by somatic cell nuclear transfer experiments in which the nuclei of oocytes were replaced with those of regular somatic cells. The resulting reprogramming of the somatic nucleus led it to adopt the function of the oocyte nuclei it had replaced, demonstrating the flexibility of the genetic information. These studies have since been demonstrated in both amphibians and mammals.[65-67] The ability to induce pluripotency in differentiated cells was further demonstrated by Yamanaka and colleagues, who identified the factors needed to reprogram cells to a pluripotent stem cell state[68] (i.e., *Oct4*, *Sox2*, *Klf4*, and *c-Myc*).

The induction of a pluripotent state in cells is known to reverse several aging hallmarks. Telomerase, the protein able to lengthen telomeres, is active in pluripotent and induced pluripotent stem (iPS) cells, in contrast to differentiated cells.[69] iPS cells are also less susceptible to DNA damage[70] and have increased mitochondrial function,[71] and iPS cells derived from senescent cells lose senescence markers and adopt the gene expression patterns of younger cells.[72]

Despite the astounding effect of these discoveries, as well as their much understood benefits in reversing aging on a cellular level, the application of cellular reprogramming to human aging has been limited.

A key issue is the act of inducing pluripotency itself. Inducing a cell to a preembryonal state invariantly forces it through a stage of embryonic-like development when trying to reproduce a differentiated cell from iPS cells. This can be dangerous, resulting in teratomas.[73-75]

The production of iPS cells is, however, a long and gradual process. Thus, it was hypothesized that shorter treatment with pluripotency reprogramming factors could help to rejuvenate cells without causing them to lose their differentiated state and functional phenotype.[76,77] Improvements in HP1β mobility, a marker of genetic age, were improved after 9 days postexposure with the reprogramming factors, although improvement returned to preexposure levels shortly afterward.[78]

To overcome this problem of phenotypic regression, a system that cyclically expressed the reprogramming factors was conceived and improved several markers of aging in a mouse model of progeria through partial iPS reprogramming. Improved hallmarks included enhancing mitochondrial function, decreasing DNA damage, and reducing cellular senescence markers.[79] This partial reprogramming, and associated improvement in aging phenotypes, was accomplished only when reprogramming factors were cyclically expressed.

Despite the progress that was made, challenges remain in translating cellular reprogramming results from in vivo and model systems to clinical treatments.

4.2 Senolytics

Cellular senescence is a major driving force of tissue aging. Senescent cells have ceased to replicate and function properly and, having become resistant to apoptosis, persist in their tissues of origin. These cells adopt a proinflammatory phenotype, the senescence-associated secretory phenotype (SASP).[80] These cells persistence in tissue and actively recruit immune cells into the tissue, which can lead to cell death, remodeling the tissue[81,82] and generally contributing to a sustained, low-grade state of global inflammation associated with aging, termed inflammaging.[83]

The detrimental effects of inflammaging, as well as the outsized effect SASP cells have on the process, make them an ideal candidate for treatment to limit the deleterious effects of aging and improve bodily health. Clearance of senescent cells by genetic ablation in mice models of aging have been shown to increase their median life span as well as improve several other general health markers.[84,85] Declines in inflammaging markers have also been noted.[86] Genetic ablation is, however, not a viable method for applying similar treatments in humans, and results only in a modest reduction in the number of total SASP cells. A more direct approach to clearing SASP cells is necessary.

This need for more direct therapies has spurred the creation of the field of senolytics, treatments designed to eliminate aged cells. The desire to eliminate specific sets of cells without harming underlying tissue means that there is a large overlap between senolytics and advances in immunotherapies against cancer. Chimeric antigen-receptor T (CAR-T) cells have proven worthy in treating B-cell lymphoma, in which CD19 CAR-T cells can be used to eliminate all $CD19^+$ B cells in lymphoma patients, allowing reconstitution of the B-cell repertoire with nonlymphoma cells.[87,88] Since these early clinical trials, cellular immunotherapies against cancer have continued to expand and diversify, but, they have not yet been successfully applied at the clinical level to the field of senolytics.

A core problem of progression in producing viable senolytics is the lack of a target. CD19 CAR-T cells were such a success in part because they are specific to CD19, a molecule whose expression is restricted to normal mature B cells, B-cell precursors, plasma cells, and the malignant B cells that characterize lymphomas.[89–92] Thus, a clear distinction is made between cells to be eliminated, CD19$^+$ cells, and all other cells. Even in this successful case, treatment has the undesired consequence of depleting a patient's normal B cells as well, compromising an arm of their immune system until it can be reconstituted. A clear marker for senescent cells has yet to be found.

Progress has been made, however, in promoting senolysis through treatment with small drug molecules. Such molecules usually inhibit proteins that discourage apoptosis. These molecules are upregulated in senescent cells, making them resistant to apoptosis.

The BCL-2 family is an important group of antiapoptotic proteins. It was shown that inhibition of BCL-2 family proteins with the flavonoid quercetin eliminated senescent cells induced by radiation, although treatment also led to thrombocytopenia.[93] The effects of quercetin were increased when it was paired with dasatinib, an inhibitor of several tyrosine kinases. The combined therapy reduced the frequency of SASP cells in white adipose tissue and the liver, increased the health span of progeroid mice, and improved several indicators of physical fitness without promoting later-life morbidity.[93,94] However, quercetin and dasatinib are both known to affect several non–BCL-2 family targets, so their observed effect may not be as specific to senescent cells as is hoped.

Another target is the inhibitor of HSP90, which stabilizes antiapoptotic proteins. Inhibition of HSP90 with the novel inhibitor 17-DMAG extended the life span in a mouse model of progeria.[95] Other possible therapeutics come from preestablished molecules. For instance, several studies showed that cardiac glycosides, inhibitors of Na^+/K^+ ATPase pumps, acted powerfully as senolytics. This was linked to further disruptions of cellular pH caused by malfunction of the pump, because SASP cells are known to have a more acidic cytosol.[96,97]

Despite the promises of senolytics, there has yet to be concerted progress in producing a clinical treatment that can be applied to human aging. Nevertheless, progress continues apace.

5. Outstanding questions and future perspectives

Advances in understanding the complicated etiology of aging have spurred similar developments in possible treatments to slow or reverse aging. Despite the many advances and the progress made, a treatment tailored specifically to aging has not yet materialized. This is not because of a lack of quality research or dedication. Rather, it hinges on some key remaining questions and mysteries that still need to be worked out. The answer, then as always, is more research.

Concrete progress has been achieved: several types of senolytics and dietary supplements designed to combat aging have made it to clinical trials, and precursors of the small molecule NAD^+ are available for purchase on the private market.

This chapter has attempted to give a broad summary of the therapeutic side of the study of aging, but it has not been able to cover all of the different branches of therapy pursued because of the vast number. Progress in the fields detailed here has been significant, but it may be that other fields will become of key interest as our understanding of the aging process deepens.

Moving forward, bioinformatics will undoubtedly have a large role in expanding the field, whether through machine learning—assisted structural studies to find more candidate small molecule drugs to inhibit important proteins involved in aging or through large screens of databases to determine cellular markers of senescence, in plasma for use as a biomarker, or on the surface for use in the immunotherapeutic removal of SASP cells. The application of bioinformatics and machine learning to produce novel drug candidates against aging has already been validated experimentally.[98]

Acknowledgments

The authors acknowledge the valuable work of the many investigators whose published articles they were unable to cite owing to space limitations. E.F.F. was supported by the National Natural Science Foundation of China (No. 81971327), the Southern and Eastern Norway Regional Health Authority (Nos. 2017056, 2020001, and 2021021), the Research Council of Norway (No. 262175), Akershus University Hospital (Nos. 269901, 261973, and 262960), the Civitan Norges Forskningsfond for Alzheimer's sykdom (No. 281931), the Czech Republic—Norway KAPPA program (with Martin Vyhnálek, No. TO01000215), and the Norwegian Cancer Society and Norwegian Breast Cancer Society (No. 207819). S.L. has received funding from the European Union's Horizon 2020 research and innovation program under the Marie Skłodowska-Curie grant agreement (No. 801133).

Competing interests

E.F.F. has a CRADA arrangement with ChromaDex (United States) and is a consultant to Aladdin Healthcare Technologies (United Kingdom and Germany), the Vancouver Dementia Prevention Center (Canada), Intellectual Labs (Norway), and MindRank AI (China).

References

1. Pontzer H, Yamada Y, Sagayama H, Ainslie Philip N, Andersen Lene F, Anderson Liam J, et al. Daily energy expenditure through the human life course. *Science.* 2021;373:808—812.
2. McCay CM, Crowell MF. Prolonging the life span. *Sci Mon.* 1934;39:405—414.
3. Vaughan KL, Kaiser T, Peaden R, Anson RM, de Cabo R, Mattison JA. Caloric restriction study Design limitations in rodent and Nonhuman primate studies. *J Gerontol: Series A.* 2018;73:48—53.
4. Mattison JA, Roth GS, Beasley TM, et al. Impact of caloric restriction on health and survival in rhesus monkeys from the NIA study. *Nature.* 2012;489:318—321.
5. Colman RJ, Beasley TM, Kemnitz JW, Johnson SC, Weindruch R, Anderson RM. Caloric restriction reduces age-related and all-cause mortality in rhesus monkeys. *Nat Commun.* 2014;5:3557.
6. Kapahi P, Kaeberlein M, Hansen M. Dietary restriction and lifespan: lessons from invertebrate models. *Ageing Res Rev.* 2017;39:3—14.
7. Walford RL, Mock D, Verdery R, MacCallum T. Calorie restriction in Biosphere 2: alterations in physiologic, hematologic, hormonal, and Biochemical parameters in humans restricted for a 2-year period. *J Gerontol: Series A.* 2002;57:B211—B224.
8. Baur JA, Pearson KJ, Price NL, et al. Resveratrol improves health and survival of mice on a high-calorie diet. *Nature.* 2006;444:337—342.
9. Longo VD, Panda S. Fasting, circadian rhythms, and time-restricted feeding in healthy lifespan. *Cell Metabol.* 2016;23:1048—1059.
10. Finkel T. The metabolic regulation of aging. *Nat Med.* 2015;21:1416—1423.
11. Goodrick CL, Ingram DK, Reynolds MA, Freeman JR, Cider N. Effects of intermittent feeding upon body weight and lifespan in inbred mice: interaction of genotype and age. *Mech Ageing Dev.* 1990;55:69—87.
12. Catenacci VA, Pan Z, Ostendorf D, et al. A randomized pilot study comparing zero-calorie alternate-day fasting to daily caloric restriction in adults with obesity. *Obesity.* 2016;24:1874—1883.
13. Hoddy KK, Kroeger CM, Trepanowski JF, Barnosky A, Bhutani S, Varady KA. Meal timing during alternate day fasting: impact on body weight and cardiovascular disease risk in obese adults. *Obesity.* 2014;22:2524—2531.
14. Varady KA, Bhutani S, Church EC, Klempel MC. Short-term modified alternate-day fasting: a novel dietary strategy for weight loss and cardioprotection in obese adults. *Am J Clin Nutr.* 2009;90:1138—1143.
15. Roberts MN, Wallace MA, Tomilov AA, et al. A ketogenic diet extends longevity and healthspan in adult mice. *Cell Metabol.* 2018;27:1156.
16. Newman JC, Covarrubias AJ, Zhao M, et al. Ketogenic diet reduces midlife mortality and improves memory in aging mice. *Cell Metabol.* 2017;26:547—557. e8.
17. Kennedy BK, Lamming DW. The mechanistic target of rapamycin: the grand ConducTOR of metabolism and aging. *Cell Metabol.* 2016;23:990—1003.

18. Wu JJ, Liu J, Chen Edmund B, et al. Increased mammalian lifespan and a segmental and tissue-specific slowing of aging after genetic reduction of mTOR expression. *Cell Rep.* 2013;4:913–920.
19. Kapahi P, Zid BM, Harper T, Koslover D, Sapin V, Benzer S. Regulation of lifespan in Drosophila by modulation of genes in the TOR signaling pathway. *Curr Biol.* 2004;14:885–890.
20. Vellai T, Takacs-Vellai K, Zhang Y, Kovacs AL, Orosz L, Müller F. Influence of TOR kinase on lifespan in *C. elegans*. *Nature.* 2003;426:620.
21. Robida-Stubbs S, Glover-Cutter K, Lamming Dudley W, et al. TOR signaling and rapamycin influence longevity by regulating SKN-1/Nrf and DAF-16/FoxO. *Cell Metabol.* 2012;15:713–724.
22. Bjedov I, Toivonen JM, Kerr F, et al. Mechanisms of life span extension by rapamycin in the fruit fly *Drosophila melanogaster*. *Cell Metabol.* 2010;11:35–46.
23. Wilkinson JE, Burmeister L, Brooks SV, et al. Rapamycin slows aging in mice. *Aging Cell.* 2012;11:675–682.
24. Lesniewski LA, Seals DR, Walker AE, et al. Dietary rapamycin supplementation reverses age-related vascular dysfunction and oxidative stress, while modulating nutrient-sensing, cell cycle, and senescence pathways. *Aging Cell.* 2017;16:17–26.
25. Chen C, Liu Y, Liu Y, Zheng P. mTOR regulation and therapeutic rejuvenation of aging hematopoietic stem cells. *Sci Signal.* 2009;2:ra75–ra.
26. Augustine JJ, Bodziak KA, Hricik DE. Use of sirolimus in solid organ transplantation. *Drugs.* 2007;67:369–391.
27. de Oliveira MA, Martins e Martins F, Wang Q, et al. Clinical presentation and management of mTOR inhibitor-associated stomatitis. *Oral Oncol.* 2011;47:998–1003.
28. Zheng Y, Collins SL, Lutz MA, et al. A role for mammalian target of rapamycin in regulating T cell activation versus anergy. *J Immunol.* 2007;178:2163.
29. Goodwin K, Viboud C, Simonsen L. Antibody response to influenza vaccination in the elderly: a quantitative review. *Vaccine.* 2006;24:1159–1169.
30. Mannick Joan B, Del Giudice G, Lattanzi M, et al. mTOR inhibition improves immune function in the elderly. *Sci Transl Med.* 2014;6, 268ra179-268ra179.
31. Mannick Joan B, Morris M, Hockey Hans-Ulrich P, et al. TORC1 inhibition enhances immune function and reduces infections in the elderly. *Sci Transl Med.* 2018;10:eaaq1564.
32. Dragowska WH, Weppler SA, Qadir MA, et al. The combination of gefitinib and RAD001 inhibits growth of HER2 overexpressing breast cancer cells and tumors irrespective of trastuzumab sensitivity. *BMC Cancer.* 2011;11:420.
33. Alshaker H, Wang Q, Böhler T, et al. Combination of RAD001 (everolimus) and docetaxel reduces prostate and breast cancer cell VEGF production and tumour vascularisation independently of sphingosine-kinase-1. *Sci Rep.* 2017;7:3493.
34. Fivenson EM, Lautrup S, Sun N, et al. Mitophagy in neurodegeneration and aging. *Neurochem Int.* 2017;109:202–209.
35. Kerr JS, Adriaanse BA, Greig NH, et al. Mitophagy and alzheimer's disease: cellular and molecular mechanisms. *Trends Neurosci.* 2017;40:151–166.
36. Lou G, Palikaras K, Lautrup S, Scheibye-Knudsen M, Tavernarakis N, Fang EF. Mitophagy and Neuroprotection. *Trends Mol Med.* 2020;26:8–20.
37. Yoshino J, Baur JA, Imai S-i. NAD+ intermediates: the Biology and therapeutic potential of NMN and NR. *Cell Metabol.* 2018;27:513–528.
38. Hikosaka K, Yaku K, Okabe K, Nakagawa T. Implications of NAD metabolism in pathophysiology and therapeutics for neurodegenerative diseases. *Nutr Neurosci.* 2021;24:371–383.
39. Rajman L, Chwalek K, Sinclair DA. Therapeutic potential of NAD-Boosting molecules: the in vivo evidence. *Cell Metabol.* 2018;27:529–547.

40. Zhu X-H, Lu M, Lee B-Y, Ugurbil K, Chen W. In vivo NAD assay reveals the intracellular NAD contents and redox state in healthy human brain and their age dependences. *Proc Natl Acad Sci U S A*. 2015;112:2876–2881.
41. Massudi H, Grant R, Braidy N, Guest J, Farnsworth B, Guillemin GJ. Age-associated changes in oxidative stress and NAD+ metabolism in human tissue. *PLoS One*. 2012;7:e42357.
42. Ramsey KM, Mills KF, Satoh A, Imai S-i. Age-associated loss of Sirt1-mediated enhancement of glucose-stimulated insulin secretion in beta cell-specific Sirt1-overexpressing (BESTO) mice. *Aging Cell*. 2008;7:78–88.
43. Fang EF, Hou Y, Lautrup S, et al. NAD+ augmentation restores mitophagy and limits accelerated aging in Werner syndrome. *Nat Commun*. 2019;10:5284.
44. Belenky P, Racette FG, Bogan KL, McClure JM, Smith JS, Brenner C. Nicotinamide riboside promotes Sir2 silencing and extends lifespan via Nrk and Urh1/pnp1/meu1 pathways to NAD+. *Cell*. 2007;129:473–484.
45. Mouchiroud L, Houtkooper Riekelt H, Moullan N, et al. The NAD+/Sirtuin pathway modulates longevity through activation of mitochondrial UPR and FOXO signaling. *Cell*. 2013;154:430–441.
46. Zhang H, Ryu D, Wu Y, et al. NAD+ repletion improves mitochondrial and stem cell function and enhances life span in mice. *Science*. 2016;352:1436–1443.
47. de Picciotto NE, Gano LB, Johnson LC, et al. Nicotinamide mononucleotide supplementation reverses vascular dysfunction and oxidative stress with aging in mice. *Aging Cell*. 2016;15:522–530.
48. Mills KF, Yoshida S, Stein LR, et al. Long-term administration of nicotinamide mononucleotide mitigates age-associated physiological decline in mice. *Cell Metabol*. 2016;24:795–806.
49. Trammell SAJ, Schmidt MS, Weidemann BJ, et al. Nicotinamide riboside is uniquely and orally bioavailable in mice and humans. *Nat Commun*. 2016;7:12948.
50. Conze DB, Crespo-Barreto J, Kruger CL. Safety assessment of nicotinamide riboside, a form of vitamin B(3). *Hum Exp Toxicol*. 2016;35:1149–1160.
51. Dellinger RW, Santos SR, Morris M, et al. Repeat dose NRPT (nicotinamide riboside and pterostilbene) increases NAD+ levels in humans safely and sustainably: a randomized, double-blind, placebo-controlled study. *Npj Aging Mech Dis*. 2017;3:17.
52. Airhart SE, Shireman LM, Risler LJ, et al. An open-label, non-randomized study of the pharmacokinetics of the nutritional supplement nicotinamide riboside (NR) and its effects on blood NAD+ levels in healthy volunteers. *PLoS One*. 2017;12:e0186459.
53. D'Amico D, Andreux PA, Valdés P, Singh A, Rinsch C, Auwerx J. Impact of the natural compound urolithin A on health, disease, and aging. *Trends Mol Med*. 2021;27:687–699.
54. González-Barrio R, Borges G, Mullen W, Crozier A. Bioavailability of anthocyanins and ellagitannins following consumption of raspberries by healthy humans and subjects with an ileostomy. *J Agric Food Chem*. 2010;58:3933–3939.
55. Tomás-Barberán FA, González-Sarrías A, García-Villalba R, et al. Urolithins, the rescue of "old" metabolites to understand a "new" concept: metabotypes as a nexus among phenolic metabolism, microbiota dysbiosis, and host health status. *Mol Nutr Food Res*. 2017;61:1500901.
56. Espín JC, Larrosa M, García-Conesa MT, Tomás-Barberán F. Biological significance of urolithins, the gut microbial ellagic Acid-derived metabolites: the evidence so far. *Evid Based Complement Alternat Med*. 2013;2013:270418.
57. Cortés-Martín A, García-Villalba R, González-Sarrías A, et al. The gut microbiota urolithin metabotypes revisited: the human metabolism of ellagic acid is mainly determined by aging. *Food Funct*. 2018;9:4100–4106.

58. Cortés-Martín A, Selma MV, Tomás-Barberán FA, González-Sarrías A, Espín JC. Where to look into the puzzle of polyphenols and health? The postbiotics and gut microbiota associated with human metabotypes. *Mol Nutr Food Res*. 2020;64:1900952.

59. Ryu D, Mouchiroud L, Andreux PA, et al. Urolithin A induces mitophagy and prolongs lifespan in *C. elegans* and increases muscle function in rodents. *Nat Med*. 2016;22:879—888.

60. Liu C-f, Li X-l, Zhang Z-l, et al. Antiaging effects of urolithin A on replicative senescent human skin fibroblasts. *Rejuvenation Res*. 2018;22:191—200.

61. Singh A, D'Amico D, Andreux PA, et al. Urolithin A improves muscle strength, exercise performance, and biomarkers of mitochondrial health in a randomized trial in middle-aged adults. *Cell Rep Med*. 2022;3:100633.

62. Chen DZ, Patel DV, Hackbarth CJ, et al. Actinonin, a naturally occurring antibacterial agent, is a potent deformylase inhibitor. *Biochemistry*. 2000;39:1256—1262.

63. Sun N, Yun J, Liu J, et al. Measuring in vivo mitophagy. *Mol Cell*. 2015;60:685—696.

64. Richter U, Lahtinen T, Marttinen P, et al. A mitochondrial ribosomal and RNA Decay pathway Blocks cell proliferation. *Curr Biol*. 2013;23:535—541.

65. Briggs R, King Thomas J. Transplantation of living nuclei from blastula cells into enucleated frogs' eggs *. *Proc Natl Acad Sci U S A*. 1952;38:455—463.

66. Gurdon JB. Adult frogs derived from the nuclei of single somatic cells. *Dev Biol*. 1962;4:256—273.

67. Wilmut I, Schnieke AE, McWhir J, Kind AJ, Campbell KHS. Viable offspring derived from fetal and adult mammalian cells. *Nature*. 1997;385:810—813.

68. Takahashi K, Yamanaka S. Induction of pluripotent stem cells from mouse embryonic and adult fibroblast cultures by defined factors. *Cell*. 2006;126:663—676.

69. Marion RM, Strati K, Li H, et al. Telomeres acquire embryonic stem cell characteristics in induced pluripotent stem cells. *Cell Stem Cell*. 2009;4:141—154.

70. Marión RM, Strati K, Li H, et al. A p53-mediated DNA damage response limits reprogramming to ensure iPS cell genomic integrity. *Nature*. 2009;460:1149—1153.

71. Suhr ST, Chang EA, Tjong J, et al. Mitochondrial rejuvenation after induced pluripotency. *PLoS One*. 2010;5:e14095.

72. Lapasset L, Milhavet O, Prieur A, et al. Rejuvenating senescent and centenarian human cells by reprogramming through the pluripotent state. *Genes Dev*. 2011;25:2248—2253.

73. Shibata H, Komura S, Yamada Y, et al. In vivo reprogramming drives Kras-induced cancer development. *Nat Commun*. 2018;9:2081.

74. Ohnishi K, Semi K, Yamamoto T, et al. Premature termination of reprogramming in vivo leads to cancer development through altered epigenetic regulation. *Cell*. 2014;156:663—677.

75. Abad M, Mosteiro L, Pantoja C, et al. Reprogramming in vivo produces teratomas and iPS cells with totipotency features. *Nature*. 2013;502:340—345.

76. Singh PB, Zacouto F. Nuclear reprogramming and epigenetic rejuvenation. *J Biosci*. 2010;35:315—319.

77. Manukyan M, Singh PB. Epigenetic rejuvenation. *Gene Cell*. 2012;17:337—343.

78. Manukyan M, Singh PB. Epigenome rejuvenation: HP1β mobility as a measure of pluripotent and senescent chromatin ground states. *Sci Rep*. 2014;4:4789.

79. Ocampo A, Reddy P, Martinez-Redondo P, et al. In vivo amelioration of age-associated hallmarks by partial reprogramming. *Cell*. 2016;167:1719—1733.e12.

80. Muñoz-Espín D, Serrano M. Cellular senescence: from physiology to pathology. *Nat Rev Mol Cell Biol*. 2014;15:482—496.

81. Yanai H, Fraifeld VE. The role of cellular senescence in aging through the prism of Koch-like criteria. *Ageing Res Rev*. 2018;41:18—33.

82. van Deursen JM. The role of senescent cells in ageing. *Nature*. 2014;509:439—446.

Canonical and novel strategies to delay or reverse aging **239**

83. Ferrucci L, Fabbri E. Inflammageing: chronic inflammation in ageing, cardiovascular disease, and frailty. *Nat Rev Cardiol.* 2018;15:505—522.
84. Baker DJ, Childs BG, Durik M, et al. Naturally occurring p16Ink4a-positive cells shorten healthy lifespan. *Nature.* 2016;530:184—189.
85. Baker DJ, Wijshake T, Tchkonia T, et al. Clearance of p16Ink4a-positive senescent cells delays ageing-associated disorders. *Nature.* 2011;479:232—236.
86. Palmer AK, Xu M, Zhu Y, et al. Targeting senescent cells alleviates obesity-induced metabolic dysfunction. *Aging Cell.* 2019;18:e12950.
87. Kochenderfer JN, Wilson WH, Janik JE, et al. Eradication of B-lineage cells and regression of lymphoma in a patient treated with autologous T cells genetically engineered to recognize CD19. *Blood.* 2010;116:4099—4102.
88. Turtle Cameron J, Hanafi L-A, Berger C, et al. Immunotherapy of non-Hodgkin's lymphoma with a defined ratio of CD8+ and CD4+ CD19-specific chimeric antigen receptor—modified T cells. *Sci Transl Med.* 2016;8, 355ra116-355ra116.
89. Nadler LM, Anderson KC, Marti G, et al. B4, a human B lymphocyte-associated antigen expressed on normal, mitogen-activated, and malignant B lymphocytes. *J Immunol.* 1983;131:244.
90. Pontvert-Delucq S, Breton-Gorius J, Schmitt C, et al. Characterization and functional analysis of adult human Bone marrow cell subsets in relation to B-lymphoid development. *Blood.* 1993;82:417—429.
91. Uckun FM, Jaszcz W, Ambrus JL, et al. Detailed studies on expression and function of CD19 surface determinant by using B43 monoclonal antibody and the clinical potential of anti-CD19 immunotoxins. *Blood.* 1988;71:13—29.
92. Harada H, Kawano MM, Huang N, et al. Phenotypic difference of normal plasma cells from mature myeloma cells. *Blood.* 1993;81:2658—2663.
93. Zhu Y, Tchkonia T, Pirtskhalava T, et al. The Achilles' heel of senescent cells: from transcriptome to senolytic drugs. *Aging Cell.* 2015;14:644—658.
94. Roos CM, Zhang B, Palmer AK, et al. Chronic senolytic treatment alleviates established vasomotor dysfunction in aged or atherosclerotic mice. *Aging Cell.* 2016;15:973—977.
95. Fuhrmann-Stroissnigg H, Ling YY, Zhao J, et al. Identification of HSP90 inhibitors as a novel class of senolytics. *Nat Commun.* 2017;8:422.
96. Guerrero A, Herranz N, Sun B, et al. Cardiac glycosides are broad-spectrum senolytics. *Nat Metabol.* 2019;1:1074—1088.
97. Triana-Martínez F, Picallos-Rabina P, Da Silva-Álvarez S, et al. Identification and characterization of Cardiac Glycosides as senolytic compounds. *Nat Commun.* 2019;10:4731.
98. Xie C, Zhuang X-X, Niu Z, et al. Amelioration of Alzheimer's disease pathology by mitophagy inducers identified via machine learning and a cross-species workflow. *Nat Biomed Eng.* 2022;6:76—93.

Index

Note: 'Page numbers followed by "*f*" indicate figures and "*t*" indicate tables.'

A

Actinonin, 230
Adaptive immunity, 212–215
Aggregated β-amyloid (Aβ) plaques, 164
Aging research, 2–3
Aging reversal/delay
  cellular reprogramming, 231–232
  nonpharmaceutical interventions, 226–228
  pharmaceutical treatments, 228–230
  prophylactic treatment, 225
  senolytics, 232–233
Aging science, 3
Amino acid misincorporation rate, protein translation, 57–58
AMPK signaling pathway, 41–43
Astrocytes, 163
Autophagosomes, 108–113, 118
Autophagy
  AMPK signaling pathway, 42–43
  autophagy-related genes (ATGs), 108–113
  chaperone-mediated autophagy (CMA), 108–113
  compromised, 115–117
  dysfunctional, 117–119
  genes, 111t–112t
  mammalian target of rapamycin (mTOR) pathway, 43
  subtypes, 108–113
  temporal and spatial effects, 107
  tissue-specific, 113–115
  types, 108–113
Autophagy–lysosome system, 65–66
Autophagy-related genes (ATGs), 108–113

B

B cells, 213–214

Blood circulation, spleen, 204
Blood endothelial cells (BECs), 206–207
Bone marrow, immunosenescence, 200–202

C

Calorie restriction (CR)
  and fasting, 227–228
  long-term, 47
  moderate, 47
  nonhuman primates, 46–47
  preclinical studies, 47
Cancer, 2–3, 219
  biology, 3
Canonical histones, 10
Cardiovascular diseases (CVDs), 219–220
Cell cycle arrest, cellular senescence, 177–178
Cellular reprogramming, aging reversal/delay, 231–232
Cellular senescence, 165–167. *See also* Senescence
Centromeric protein A (CENP-A), 18–19
Chaperome, 58
Chaperone-mediated autophagy (CMA), 108–113
Chaperone-mediated protein quality control, 76–78
Chaperones linked to protein synthesis (CLIPS), 58–59
Chemokines, 157
  CXC-chemokine ligand 12 (CXCL12), 201–202
Chimeric antigen-receptor T (CAR-T) cells, 232
Chromatin structure, 9

242 Index

Chromatin structure (*Continued*)
  accessibility changes, 14–15
  age-associated heterochromatin loss, 14
  nucleosome remodeling, 14–15
Clearance module, proteostasis
  asymmetric cell divisions, 67
  autophagy–lysosome system, 65–66
  degradation module, 63
  fibroblasts, 67–68
  misfolded protein secretion, 68
  terminal sequestration, 66–67
  ubiquitin–proteasome system, 63–65
Common lymphoid progenitor (CLP)
    cells, 201
Common myeloid progenitor (CMP)
    cells, 201
Compromised mitophagy, 125–126,
    168–171
Constitutive heterochromatin, 9
COVID-19 pandemic, 2
Cytokines, 157

D
Dendritic cells (DCs), 212
De novo folding, 59–60
Deoxyribonucleic acid modifications, 11
De-ubiquitylases (DUBs), 64
DNA methyltransferase (DNMT), 11
Dynamin-related protein 1 (Drp1)-
    mediated mitochondrial fission,
    116–117
Dysfunctional autophagy, 117–119

E
Electron transport chain (ETC),
    119–121
Endoplasmic reticulum (ER)–associated
    degradation (ERAD), 65
Epigenetics
  alterations, 28f
  chromatin structure, 9, 14–15
  deoxyribonucleic acid modifications, 11
    methylation changes, 23–24
    methylation clocks and predictors,
      24–26
  histone
    expression levels, 16

posttranslational modifications,
      10–11, 19–23
  variants, 10, 16–19
human aging research models, 12–14
induced pluripotent stem cells (iPSCs)
    generation, 27
molecular mechanisms, 9
noncoding ribonucleic acids (ncRNAs),
    12, 26
regulatory mechanisms, 9
rejuvenation, 27
reversal, 27
Euchromatin, 9
Eukaryotic proteostasis
  protein complex assembly, 69–70
  targeting, 68–69
Executive modules, proteostasis network
  dysregulation, 80–83
  organism-wide proteostasis control, 72
  stress response, 70–72
Extracellular insoluble molecules, 158
Extracellular proteases, 157–158

F
Facultative heterochromatin, 9
Fibroblastic reticular cells (FRCs), 207
Fibronectin, 158
Folding module, proteostasis, 58–62
  ATP-coupled unfolding, 58
  chaperone-assisted folding, 59f
  chaperone networks, 58–59
  conformational maintenance, 60–61
  de novo folding, 59–60
  misfolded protein triage, 60–61
  spatial sequestration and resolution,
    61–62
  three-dimensional (3D) structure, 58
Follicular dendritic cells (FDCs),
    207–208
Forkhead box transcription factors
    (FOXOs), 45–46

G
Genomic instability, cellular senescence,
    165–167
Genomic process regulation, 9
Glycophagy, 109t

Index **243**

Growth hormone (GH) stimulation, 45–46

H

Hallmarks, aging, 4–5

Healthspan, 3–4

Heat shock protein (HSP), 58–59, 61

Hematopoietic stem cells (HSCs)
 aging-associated increase, 200
 altered homing and mobilization, 201–202
 skewed differentiation potential, 201

Heterochromatin, 9

High endothelial venule (HEV), 206–207

Histone
 expression level changes, 16
 posttranslational modifications, 10–11
  acetylation, 21–23
  chromatin marks, 19–20
  H3K9 methylation marks, 11
  H3K27me3 repressive mark, 19–20
  methylation, 20
  principal component analysis, 19–20
 variants, 10, 17t
  2A variants, 16–18
  H2AJ, 17
  H2AX variant, 16–17
  H3 variant, 18–19
  macroH2A (mH2A) histone, 17–18

Human aging research models, 12–14

Hutchinson—Gilford progeria syndrome (HGPS), 12–13

I

Immunity
 adaptive, 212–215
 innate, 208–212

Immunosenescence
 and age-associated diseases
  cancer, 219
  cardiovascular diseases (CVDs), 219–220
  metabolic diseases, 220–221
  neurodegenerative diseases, 217–218
  rheumatoid arthritis, 218–219
 and aging process

cellular senescence and immunosuppression, 216–217
  immunosuppressive network, 215–216
  inflammaging, 215–216
 complementary process, 199
 lymphoid organs
  bone marrow, 200–202
  lymph nodes (LNs), 205–208
  spleen, 204–205
  thymus, 202–204

Immunosuppressive network, 215–216

Inflammaging, 215–216

Inflammatory interleukins (ILs), 156–157

inhibitory mechanism, IAMPK signaling pathway, 43

Innate immunity, 208–212

Insoluble protein deposit (IPOD), 66–67

Insulin-like growth factor (IGF), 45–46, 157

Intermittent fasting (IF), 227

L

Lifespan, 3–4

Lipofuscin (LF), 178–179

Lipophagy, 109t

Long noncoding RNAs (lncRNAs), 12

Loss of protein complex stoichiometry, 81–82

Lymph nodes (LNs)
 blood endothelial cells, 206–207
 chronic and progressive changes, 205–206
 fibroblastic reticular cells, 207
 follicular dendritic cells (FDCs), 207–208
 high endothelial venule cells, 206–207
 lymphatic vessels and lymphatic endothelial cells, 206
 stromal cells, 205

Lysophagy, 109t

M

Macroautophagy, 66, 108–113
 core machinery, 110f

244 Index

Macroautophagy (*Continued*)
and senescence, 170
MacroH2A (mH2A) histone, 17—18
Macrophages, 211—212
Mammalian target of rapamycin
(mTOR), 43—44, 228
Matrix metalloprotease (MMP),
158
Mesenchymal stem cells (MSCs), senescence, 175—176
Metabolic diseases, 220—221
Methylation
deoxyribonucleic acid, 23—26
histone, 20
Microautophagy, 108—113
Microglia, 163
Microglia-dependent phagocytosis,
115—116
Mitochondria
bioenergetics, 121
biology, 119—121
structures, 120f
Mitochondrial deoxyribonucleic acid
(mtDNA), 119—121
mutations in, 127
Mitochondrial dysfunction
apoptosis and inflammatory reactions,
134—136
energy shortages and aging cells,
132—136
high level reactive oxygen species,
129—130
low level nicotinamide adenine dinucleotide, 130—132
mechanisms, 122—132, 133f
oxidative phosphorylation dysfunctions,
128—129
reduced mitochondrial biogenesis and
integrity, 122—123
reduced mitochondrial dynamics,
123—125
Mitochondrial dysfunction-induced senescence (MiDAS), 154, 169—170
Mitochondrial membrane potential
(MMP) changes, 124—125
Mitofusin 1 and 2 (MFN1 and MFN2),
123—124
Mitophagy, 109t
and autophagy

compromised, 115—117
dysfunctional, 117—119
tissue-specific, 113—115
inducers, 229—230
pathways, 113, 114f
and senescence, 170—171
Monocytes, 210—212
Myeloid dendritic cells (mDCs),
212
Myeloid-derived suppressor cells
(MDSCs), 215—216

N
Natural killer (NK) cells, 209—210
Neurodegeneration, 163—164
Neurodegenerative diseases,
217—218
Neurofibrillary tangles (NFTs), 164
Neutrophil extracellular traps (NETs),
208—209
Neutrophils, 208—209
Nicotinamide adenine dinucleotide
(NAD)
deficiency, 44
low level, 130—132
Nicotinamide riboside (NR), 45
Noncoding ribonucleic acids (ncRNAs),
12, 26
Nonpharmaceutical interventions
calorie restriction and fasting,
227—228
folk treatments, 226
Nonprotein secretion, 158—159
Nuclear changes, cellular senescence,
178
Nucleophagy, 109t
Nucleosome remodeling, 14—15
Nucleosome remodeling and deacetylase
(NURD) complex, 15
Nutrient sensing and aging
AMPK signaling pathway, 41—43
calorie restriction, 46—47
cellular energy supply, 41
insulin-like growth factor 1 (IGF-1)
pathway, 45—46
mammalian target of rapamycin
(mTOR) pathway, 43—44
nutrient deficiency, 41
sirtuin pathway, 44—45

Index   245

O

Oligodendrocyte progenitor cells (OPC), 164
Oncogene-induced senescence (OIS), 171—172
Organism-wide proteostasis control, 72
Oxidized nicotine adenine dinucleotide (NAD+), 229—230

P

Pathogen identification, 2—3
Pharmaceutical treatments
  mitophagy inducers, 229—230
  rapamycin and rapalogs, 228—229
  senolytics, 232—233
p16-induced senescence, 151—153
p53-induced senescence, 151—153
Plasmacytoid dendritic cells (pDCs), 212
Pluripotent stem cells (PSCs), 175
Pneumonia, 4
Posttranslational modifications, histone, 10—11
Progeroid syndromes (PS), 12—13
Protein targeting defects, 80—81
Proteome turnover, 76
Proteostasis network
  functional modules, 56—72, 57f
    auxiliary protein, 56
    clearance module, 62—68
    dysregulation, 74—80, 75f
    executive modules, 70—72
    extended modules, 68—70
    folding module, 58—62
    synthesis module, 56—58
  remodeling, 73—83
Proteotoxic stress responses, 71—72
PTEN-induced putative kinase 1 (PINK1)—parkin pathway of mitophagy, 114f

Q

Q-bodies, 61—62

R

Rapamycin and rapalogs, 228—229
Receptor-mediated mitophagy pathways, 114f

Replication-independent histones, 10
Replicative aging model, 16
Repressive marks, histone modifications, 10—11
Reticulophagy, 109t
Rheumatoid arthritis, 218—219

S

Secretory phenotype, cellular senescence, 179
Senescence, 12—13
  acute, 165
  age-dependent senescence, 166f—167f
  age-related diseases, 163
  cell characteristics, 149—151
  cellular, 165—167
  compromised mitophagy, 168—171
  features, 150f
  fibroblasts, 159
  gene and protein, 152t
  genomic instability, 165—167
  impaired mitochondrial function, 168—171
  in vitro, 177—180
  in vivo, 180
  mouse models, 159—163, 161t—162t
  and neurodegeneration, 163—164
  oncogene-induced, 171—172
  p53 and p16, 151—153
  senotherapies, 176—177
  and stem cell therapy, 175—176
  telomere attrition, 167—168
Senescence-associated heterochromatin foci (SAHFs), 14, 178
Senescence-associated signaling phenotype (SASP), 149—151, 232—233
  cell types, 154
  chromatin alterations, 154
  constituents, 154
  extracellular insoluble molecules, 158
  extracellular proteases, 157—158
  soluble signaling factors, 154—157
    chemokines, 157
    insulin-like growth factor—binding proteins, 157
    proinflammatory interleukins and cytokines, 154—157

Senolytics, 176—177, 232—233
Senomorphics, 176—177
Senotherapies, 176—177
Sirtuin pathway
  acetyl group removal, 44
  mammalian SIRT1—7, 44—45
  nicotinamide adenine dinucleotide
    (NAD) deficiency, 44
  nicotinamide mononucleotide (NMN)
    supplementation, 45
  SIRT1 and SIRT6, 44—45
  vascular aging, 44—45
Small heat shock protein (sHSP), 61—62
Spatial compartmentalization, 80—81
Spleen, immunosenescence, 204—205
Stem cell exhaustion, 173—176
Stress resilience delice, 82—83
Stress response, proteostasis network,
  70—72
Synthesis module, proteostasis, 56—58

T
T cells, 214—215
T-cell senescence, 217—218

Telomerase, 231
Telomere
  attrition, 167—168
  damage, 151—153
Thymus
  involution, 202—204
  location, 202
Thymus epithelial cell (TEC), 203—204
Tissue-specific autophagy, 113—115
Transcription factor EB (TFEB),
  108—113
Type 2 diabetes (T2D), 220—221

U
Ubiquitin—proteasome system, 63—65
Urokinase-type PA (uPA), 158
Urolithin A (UA), 117, 230

W
Werner syndrome (WS), 12—13

X
Xenophagy, 109t

Printed in the United States
by Baker & Taylor Publisher Services